Proceedings of the
3rd Annual RECOMB Workshop

REGULATORY GENOMICS

SERIES ON ADVANCES IN BIOINFORMATICS AND COMPUTATIONAL BIOLOGY

ISSN: 1751-6404

Vol. 1: Proceedings of the 3rd Asia-Pacific Bioinformatics Conference
Eds: Yi-Ping Phoebe Chen and Limsoon Wong

Vol. 2: Information Processing and Living Systems
Eds: Vladimir B. Bajic and Tan Tin Wee

Vol. 3: Proceedings of the 4th Asia-Pacific Bioinformatics Conference
Eds: Tao Jiang, Ueng-Cheng Yang, Yi-Ping Phoebe Chen and Limsoon Wong

Vol. 4: Computational Systems Bioinformatics 2006
Eds: Peter Markstein and Ying Xu
ISSN: 1762-7791

Vol. 5: Proceedings of the 5th Asia-Pacific Bioinformatics Conference
Eds: David Sankoff, Lusheng Wang and Francis Chin

Vol. 6: Proceedings of the 6th Asia-Pacific Bioinformatics Conference
Eds: Alvis Brazma, Satoru Miyano and Tatsuya Akutsu

Vol. 7: Computational Methods for Understanding Bacterial and Archaeal Genomes
Eds: Ying Xu and J. Peter Gogarten

Vol. 8 Regulatory Genomics
Eds: Leong Hon Wai, Sung Wing-Kin and Eleazar Eskin

Series on Advances in Bioinformatics and Computational Biology – Volume 8

Proceedings of the
3rd Annual RECOMB Workshop

REGULATORY GENOMICS

National University of Singapore, Singapore

17 – 18 July 2006

editors

Leong Hon Wai
National University of Singapore
Singapore

Sung Wing-Kin
National University of Singapore
Singapore

Eleazar Eskin
University of California
Los Angeles, USA

ICP Imperial College Press

Published by

Imperial College Press
57 Shelton Street
Covent Garden
London WC2H 9HE

Distributed by

World Scientific Publishing Co. Pte. Ltd.
5 Toh Tuck Link, Singapore 596224
USA office: 27 Warren Street, Suite 401-402, Hackensack, NJ 07601
UK office: 57 Shelton Street, Covent Garden, London WC2H 9HE

British Library Cataloguing-in-Publication Data
A catalogue record for this book is available from the British Library.

Series on Advances in Bioinformatics and Computational Biology — Vol. 8
REGULATORY GENOMICS
Proceedings of the 3rd Annual RECOMB Workshop

ISBN-13 978-1-84816-251-8
ISBN-10 1-84816-251-0

Printed in Singapore.

FOREWORD

The RECOMB Regulatory Genomics workshop continues the tradition of previous RECOMB satellite meetings, which are focussed workshops on topics of particular interest to the computational biology community – in this case, Regulatory Genomics. The workshop is an annual forum that brings together the diverse group of researchers in the field – with talks from leading researchers and submitted papers and posters on latest research results. The attendees will also be able to take the opportunity to share research results and exchange ideas. We hope that many collaborations will be born among the attendees.

The Third RECOMB Regulatory Genomics workshop will be held in Singapore at the National University of Singapore (NUS) on 17–18 July 2006. The workshop is organised as one of the activities within a two-month *Program of Algorithmic Biology: Algorithmic Techniques in Computational Biology* held in June–July 2006 at the Institute for Mathematical Sciences (IMS) of the National University of Singapore. The program also includes a *Workshop on BioAlgorithmics* (July 12–14), two tutorials (July 11 and 21), a Math-CS Camp for high school students (July 19), and a public lecture on *The Role of Mathematics and Computer Science in Computational Biology* (July 19).

The organisers of the workshop would like to express their gratitude to the invited speakers and the authors of the submitted papers and posters, the members of the programme committee and the additional external reviewers, members of the organising and the local arrangement committees, and the participants for their respective contributions to this workshop.

Finally, the Third RECOMB Satellite Workshop on Regulatory Genomics would not have been possible with the generous support of the following: the National University of Singapore – Institute for Mathematical Sciences (IMS), the School of Computing (SOC), and the Office of Life Sciences (OLS); the Association for Medical and Bio-Informatics, Singapore (AMBIS); and the Genome Institute of Singapore (GIS).

Eleazar Eskin
Hon Wai Leong
Pavel Pevzner
Wing-Kin Sung

RECOMB REGULATORY GENOMICS ORGANIZATION

RECOMB Regulatory Genomics Steering Committee

Pierre Baldi	University of California, Irvine
Michael Eisen	Lawrence Berkeley National Lab
Eleazar Eskin*	University of California, San Diego
Pavel Pevzner	University of California, San Diego

Program Committee

Mathieu Blanchette	McGill University
Guillaume Bourque	Genome Institute of Singapore
Martha Bulyk	Harvard Medical School
Gal Chechik	Stanford University
Francis Chin	University of Hong Kong
Eleazar Eskin*	University of California, San Diego
Irit Gat-Viks	Tel Aviv University
Mikhail Gelfand	Moscow State University
Sridhar Hannenhalli	University of Pennsylvania
Uri Keich	Cornell University
Tak Wah Lam	University of Hong Kong
Christina Leslie	Columbia University
Hon Wai Leong	National University of Singapore
Hao Li	University of California, San Francisco
See-Kiong Ng	Institute for Infocomm Research
Yitzhak Pilpel	Weizmann Institute of Science
Zhaohui Steve Qin	University of Michigan
Ben Raphael	University of California, San Diego
Mireille Regnier	INRIA
Marie-France Sagot	INRIA
Wing-Kin Sung*	National University of Singapore
Eran Segal	Weizmann Institute of Science
Amos Tanay	Tel Aviv University
Limsoon Wong	National University of Singapore
Christopher Workman	University of California, San Diego
Eric Xing	University of California, Berkeley

*committee chair

Organizing Committee

Francis Chin	University of Hong Kong
Hon Wai Leong*	National University of Singapore
Wing-Kin Sung	National University of Singapore
Limsoon Wong	National University of Singapore
Louxin Zhang	National University of Singapore
Zhiyong Zhang	National University of Singapore

Local Arrangement Committee

Siang Yong Yap	School of Computing, NUS
Theresa Koh	School of Computing, NUS
Lay Khim Chng	School of Computing, NUS
Noraiszah Hamzah	School of Computing, NUS
Alexia Yoke Yee Leong	School of Computing, NUS
Stephen Choon Yeow Wee	School of Computing, NUS
Wai Kin Leong	School of Computing, NUS
Philip Hui Lip Lim	School of Computing, NUS
Wendy Pei See Tan	Institute for Mathematical Sciences, NUS
Claire Li Fong Tan	Institute for Mathematical Sciences, NUS
Jolyn Wai Yeng Wong	Institute for Mathematical Sciences, NUS
Emily Ee Cheng Chan	Institute for Mathematical Sciences, NUS

CONTENTS

Contributed Papers

Keynote Papers

COMPUTATIONAL PREDICTION OF REGULATORY ELEMENTS BY COMPARATIVE SEQUENCE ANALYSIS

MARTIN TOMPA
University of Washington

With many vertebrate genomes now completely sequenced, the most promising methods for predicting functional sequence elements are based on comparison of sequences from multiple species. We focus on problems that arise when using such tools on a genome-wide scale in the vertebrates. These problems include difficulties in finding reliably homologous promoter sequences, difficulties in choosing the best tool and parameters to apply to these sequences, and difficulties in assessing the significance of the predictions produced. Solutions are offered to each of these problems, though they are far from complete.

A TALE OF TWO TOPICS — MOTIF SIGNIFICANCE AND SENSITIVITY OF SPACED SEEDS

MING LI

University of Waterloo

Computing the p-value of a motif has been a very difficult problem. Many heuristic algorihms try to approximate it. It turns out that this problem is very similar to the optimal spaced seed design in homology search. Connecting the two topics, for the first time we show computing the p-value is NP-hard, and give a reasonably fast algorithm by dynamic programming. Test results will be given.[a]

[a]Joint work with J. Zhang, Bo Jiang, J. Tromp, X. Zhang, and M.Q. Zhang.

COMPUTATIONAL CHALLENGES FOR TOP-DOWN MODELING AND SIMULATION OF BIOLOGICAL PATHWAYS

SATORU MIYANO

Human Genome Center, Institute of Medical Science, University of Tokyo

If the concept of ordinary/partial differential equations would be the only way for modeling biological pathways for simulation, our understanding of life as system through computation would be not be drastically increased and would be very biased. If the language for modeling and describing biological pathways would not be rich, we would lose a lot of valuable knowledge and information on biological systems produced and reported. Placing this understanding as our basis of development, we have been developing an XML format Cell System Markup Language CSML (http://www.csml.org/) and a modeling and simulation tool Cell Illustrator (http://www.gene-networks.com/). In this talk, we present the newest version CSML 3.0 and Cell Illustrator 3.0 which supports CSML 3.0.

Cell Illustrator (CI for short) is a software tool for modeling and simulating biological pathways which is based on the notion of Petri net which was developed with the name Genomic Object Net [1]. An important challenge for Systems Biology is to create a software platform with which scientists in biology/medicine can comfortably create models of dynamic causal interactions and processes in the cell(s) and simulate them for further investigations, e.g. testing/creating hypotheses. CI employs the notion of Hybrid Functional Petri Net with extension (HFPNe) as its architecture [2]. HFPNe was defined by enhancing some functions to hybrid Petri net so that various aspects in pathways can be intuitively modeled, including integer, real, string, boolean, vector, objects, etc. The architecture of CI 3.0 is designed so that users can get involved with modeling and simulation in a biologically intuitive way with their profound knowledge and insights, and they can also be benefited from some public/commercial pathway databases. Its effectiveness has been demonstrated by modeling various biological processes. Recently, we have developed a method for automatic parameter estimation for HFPN models by developing a theory of data assimilation that will be implemented as a function of CI.

Some XML formats have been proposed to be standard formats for biological pathways. However, all formats provide only a partial solution for the storage and integration of biological data. The aim of CSML 3.0 is to create a really usable XML format for visualizing, modeling and simulating biological pathways. Other XML formats, SBML 2.0 and CellML 1.0 are proposed and developed for dynamic simulation. These formats have become popular for chemical reactions and many applications support them as data exchanging formats. However, these formats do not define any graphical elements, which cause a difficulty to be a powerful data exchange format among biological pathway applications. Here, CSML 3.0 is developed as an integrated/unified data exchange format which covers widely used data formats and applications, e.g. CellML 1.0, SBML 2.0, BioPAX, and Cytoscape.

We also developed automatic conversion programs which convert SBML 2.0 to CSML 3.0 and CellML 1.0 to CSML 3.0 automatically. CI 3.0 fully supports CSML 3.0 as its base XML. Thus every model in SBML 2.0 and CellML 1.0 is executable on CI 3.0. It is also possible to automatically convert KEGG and BioCyc metabolic pathways to CSML.

1. Genomic Object Net: http://www.genomicobject.net/
2. Nagasaki, M., Doi, A., Matsuno, H., Miyano, S. Computational modeling of biological processes with Petri net based architecture. In "Bioinformatics Technologies" (Y.P. Chen, ed). Springer Press. 179-243, 2005.

AN IMPROVED GIBBS SAMPLING METHOD FOR MOTIF DISCOVERY VIA SEQUENCE WEIGHTING

TAO JIANG

University of California at Riverside

The discovery of motifs in DNA sequences remains a fundamental and challenging problem in computational molecular biology and regulatory genomics, although a large number of computational methods have been proposed in the past decade. Among these methods, the Gibbs sampling strategy has shown great promise and is routinely used for finding regulatory motif elements in the promoter regions of co-expressed genes. In this paper, we present an enhancement to the Gibbs sampling method when the expression data of the concerned genes is given. A sequence weighting scheme is proposed by explicitly taking gene expression variation into account in Gibbs sampling. That is, every putative motif element is assigned a weight proportional to the fold change in the expression level of its downstream gene under a single experimental condition, and a position specific scoring matrix (PSSM) is estimated from these weighted putative motif elements. Such an estimated PSSM might represent a more accurate motif model since motif elements with dramatic fold changes in gene expression are more likely to represent true motifs. This weighted Gibbs sampling method has been implemented and successfully tested on both simulated and biological sequence data. Our experimental results demonstrate that the use of sequence weighting has a profound impact on the performance of a Gibbs motif sampling algorithm.[a]

[a]Joint work with Xin Chen (School of Physical and Mathematical Sciences, Nanyang Technological University, Singapore).

DISCOVERING MOTIFS WITH TRANSCRIPTION FACTOR DOMAIN KNOWLEDGE

FRANCIS CHIN
Hong Kong University

Finding the binding sites of transcription factors from a set of promoter regions of co-regulated genes is an important problem in molecular biology. Most motif-discovering algorithms consider over-represented similar patterns as binding sites and find the position specific score matrix (PSSM) with the maximum likelihood as the solution motif. However, many motifs in real biological data cannot be discovered by these algorithms because they do not consider the biological characteristics of binding sites. We introduce a new algorithm, DIMDom, which exploits two kinds of information: (a) the characteristic pat-tern of binding site classes, where class is determined based on biological information about transcription factor domains and (b) posterior probabilities of these classes. We compared the performance of DIMDom with MEME on all the transcription factors of Drosophiia in the TRANSFAC database and found that DIMDom outperformed MEME with more than double the number of successes and double the accuracy in finding binding sites and motifs.[a]

[a]Joint work with Henry Leung.

APPLICATIONS OF ILP IN COMPUTATIONAL BIOLOGY

ANDREAS DRESS
Shanghai Institutes for Biological Sciences

In the lecture, I will present various problems in Computational Biology related in particular to phylogenetic combinatorics and network analysis that can be approached successfully using Integer Linear Programming.

ON THE EVOLUTION OF TRANSCRIPTION REGULATION NETWORKS

RON SHAMIR

Tel Aviv University

We are developing methods that employ sequence, expression and other data from multiple species, in order to identify transcription factor-DNA interactions and to trace their evolution. I will discuss several of our efforts in this direction:

- A study on the dynamics of minute changes in the regulatory sequences, using genomes of four closely related yeast species.
- Analysis of stability and change in transcriptional modules in 17 yeast species, using expression and sequence data.
- An integrated genome-wide evolutionary model of the regulatory code.

The emerging picture of the evolution in transcription regulation networks is quite fascinating. If time allows, I will also discuss our novel software tool for large scale de novo motif finding.[a]

[a]Joint work, in parts, with Amos Tanay (Rockefeller), Irit Gat-Viks, Chaim Linhart, Yonit Halperin, Daniela Raijman, (TAU) and Aviv Regev (Harvard).

SYSTEMS PHARMACOLOGY IN CANCER THERAPEUTICS: ITERATIVE INFORMATICS-EXPERIMENTAL INTERFACE

EDISON LIU

Genome Institute of Singapore

Systems biology, as a discipline, seeks to explain biologic phenomenon through the net interactions of all cellular and biochemical components within a cell or organism. We present work that uses a systems approach to build the framework for predictive pharmacology. We used as a model system, the p53 transcriptional response in vitro and in human tumors. First, we analyzed transcript profiles In 251 primary breast cancers in which the p53 gene had been sequenced and identified a clinically embedded 32-gene expression signature that distinguishes p53-mutant and wild-type tumors of different histologies that outperforms p53 sequencing. Thus, the transcriptional fingerprint is a more definitive downstream indicator of p53 function. Second, we identified a unique role for glycogen synthesis kinase-3beta (GSK-3beta) in regulating p53 function in human colorectal cancer cells. Pharmacologic modulation of GSK-3beta markedly impaired p53-dependent transactivation of targets including p21 and Puma but promoted p53-dependent conformational activation of Bax leading to apoptosis. Thus, the cell cycle arrest after p53-mediated damage response is converted to apoptosis following exposure to a variety of chemotherapeutic agents (Tan, et al. Cancer Res. 65(19):9012-20., 2005). The success of this compound will depend on a reliable assessment of p53 status in primary tumors. Based on these observations, we sought to identify the precise mechanisms of p53 gene regulation by developing a robust approach that couples chromatin immunoprecipitation (ChIP) with the paired-end ditag (PET) sequencing strategy for unbiased and precise global localization of p53 binding sites. From a saturated sampling of over half a million PET sequences, we characterized 65,572 unique p53 ChIP DNA fragments and established overlapping PET clusters as a readout to define p53 binding loci with remarkable specificity. Based on this information, we refined the consensus p53 binding motif, identified at least 542 binding loci with high confidence, discovered 98 previously unidentified p53 target genes that were implicated in novel aspects of p53 functions such as cell adhesion and motility. Finally, we showed their clinical relevance to p53-dependent tumorigenesis in primary cancer samples (Wei CL, et al. Cell. 124(1):207-19, 2006). The mutually supporting discovery framework we have established at the GIS has been the key to maximally exploiting individual discoveries in a collective manner.

COMPUTATIONAL STRUCTURAL PROTEOMICS AND INHIBITOR DISCOVERY

RUBEN ABAGYAN

The Scripps Research Institute

Rapid advance of structural proteomics calls for the development of new methods for predicting structural changes, association, function, as well as improving methods for structure based molecular design. The main challenges of computational structural biology and chemistry will be reviewed. We have developed methods for predicting the functional map of a protein with a known 3D structure, accurate docking of compounds to a binding site and virtual ligand screening of large chemical databases, and structure prediction by global energy optimization, e.g. characterizing mutants and SNPs, homology modeling, protein protein or peptide docking, and accurate loop prediction.

Predicting how flexible molecules dock to a flexible receptor is one of the main challenges in computational structural biology and structure based ligand design. Two stories in which novel compounds were discovered through "ligand-guided" receptor pocket modeling followed by virtual screening of large compound libraries, were presented. First, we developed models of the androgen receptor in an antagonist-bound conformation. These models were used to discover computationally the secondary activity of antipsychotic drugs. These drugs were then chemically altered and "re-purposed" to loose their binding to the serotonin and dopamin receptors, and improve their anti-androgen properties. The experimental side of this project was performed by the labs of Xiaokun Zhang and James Dalton. Second, in a collaboration with the David Lomas lab at Cambridge, we identified the first small molecules to inhibit pathological polymerization of an alpha1-antitrypsin mutant which is the most common genetic cause of a lethal liver disease in childhood. Computationally this project was particularly difficult because the target of a small molecule was a dynamic protein-protein interface. Third, we developed a protocol for protein-protein docking which produced the winning overall predictions in two consecutive CAPRI competitions.

Finally, a new way to disseminate structural and functional information in structural proteomics developed in collaboration with the Oxford Center for Structural Genomics is presented.

CHARACTERIZATION OF TRANSCRIPTIONAL RESPONSES TO ENVIRONMENTAL STRESS BY DIFFERENTIAL LOCATION ANALYSIS

HAIXU TANG

Indiana University

Unicellular organisms like yeast, need to rapidly respond to environmental condition changes for their survival. Using high-throughput location analysis (Chromatin Immuno-Precipitation on DNA chip, or ChIP-Chip in short), Harbison et al. have determined the genomic binding locations of 204 transcription factors (TFs) from the yeast Saccharomyces cerevisiae in rich media condition and 13 stress conditions. Here, we report a statistical method for differential location analysis, to determine the set of regulators that bind to significantly different genomic regions under certain stress conditions. From the published ChIP-Chip data by Harbison et al., we were able to identify 105 TFs-condition pairs which showed statistically significant differential binding patterns ($p < 0.05$). Comparison with published Microarray data revealed that the expression levels of nearly half of the tested TFs did not significantly change under the corresponding environmental stress, which implies that such regulatory responses would not be revealed solely by Microarray data. In conclusion, complementary differential analyses (e.g. differential location analysis) are required, in addition to commonly used Microarray-based differential expression analysis, in order to understand the global picture of cellular responses to environmental stresses.

A KNOWLEDGE-BASED HYBRID ALGORITHM FOR PROTEIN SECONDARY STRUCTURE PREDICTION

HSU WEN LIAN

Institute of Information Sciences, Academia Sinica

In our previous approach, we proposed a hybrid method called HYPROSP II for protein secondary structure prediction, which combined our proposed knowledge-based prediction algorithm PROSP and a neural net approach PSIPRED. In this talk, we further improve the performance of PROSP by proposing a better voting strategy and a wider coverage rate using both 7-mers and 5-mers.

Generally speaking, the knowledge-base algorithm, PROSP, does not necessarily provide the best result among all secondary structure prediction systems, restricted by its coverage rate. Therefore, we need to consult other algorithms or biological properties that are potentially complementary to those of PROSP. We will illustrate a neural network model to help us make good combined results from more than one system, which could be substantially better than any single system. Our approach provides a general platform of knowledge-based approach for prediction algorithms, which is more amenable to various biological domain knowledge.[a]

[a]Joint work with Ei-Wen Yang, Jia-Ming Chang, Hsin-Nan Lin, and Ting-Yi Sung.

MONOTONY AND SURPRISE
(CONSERVATIVE APPROACHES TO PATTERN DISCOVERY)

ALBERTO APOSTOLICO

Georgia Institue of Technology

Pattern discovery is often torn between the rigidity of the model and the abundance of candidates, a circumstance that tends to generate daunting computational burdens, and to give rise to a throughput that is impossible to visualize and digest.

While part of these problems is endemic, another part seems rooted in the characterizations traditionally offered for the notion of a motif or association, that are typically based either on syntax or on statistics alone. This talk describes alternate notions based on constraints of saturation that tightly combine syntactic and statistical specifications, and shows how they afford significant parsimony in the generation and testing of candidate patterns.

EVOLUTION OF BACTERIAL REGULATORY SYSTEMS

MIKHIAL S. GELFAND
Institute for Information Transmission Problems

Comparative analysis of bacterial genomes allows not only for identification of new regulatory systems and functional annotation of hypothetical genes, but also for characterization of changes in regulatory patterns. Although it is premature to speak about a theory of regulatory evolution, some patterns start to emerge. I will present results of genomic analysis of several systems of varying complexity. In particular, I will show how computational analysis of NrdR, a universal regulator of ribonucleotide reductases, has resulted in a detailed description of the regulatory signal and the mechanism of regulation, and has established links between this regulon and replication. I will present examples of regulon expansion, contraction, merging and disappearance in the metabolic pathways of oligosaccharide and sugar utilisation. Finally, I will attempt to reconstruct the evolutionary history of the regulation of iron homeostasis system in alpha-proteobacteria.

Contributed Papers

TSCAN: A TWO-STEP *DE NOVO* MOTIF DISCOVERY METHOD

O. ABUL*
Department of Cancer Research and Molecular Medicine,
Norwegian University of Science and Technology,
Trondheim, Norway
E-mail: osman.abul@ntnu.no

G. K. SANDVE
Department of Computer and Information Science,
Norwegian University of Science and Technology,
Trondheim, Norway
E-mail: sandve@idi.ntnu.no

F. DRABLØS
Department of Cancer Research and Molecular Medicine,
Norwegian University of Science and Technology,
Trondheim, Norway
E-mail: finn.drablos@ntnu.no

Computational discovery of novel motifs in biological sequences is an important and well-studied problem. The key to motif discovery methods, either *de novo* or library based, is having well-defined scoring functions. Several different scalar valued scoring functions have been proposed that measure some property of biological motifs. However, there is no general scoring function capable of unifying all properties together. In this work, we propose a two-step *de novo* motif discovery paradigm employing two scoring functions measuring different properties of biological relevance. We define a word-counting based method, called *TScan*, using this paradigm. It is mainly inspired by *MDScan*, but does not require supplementary ChIP-chip data. Our results on seven data sets from a recent study are promising, with discovered motifs agreeing well with the consensus motifs defined for the data sets.

1. Introduction

Traditionally, motif discovery algorithms fall into two categories: *de novo* methods and library methods. Methods in the former category (*e.g. MEME* [2], *MDScan* [10], *Bioprospector* [9], *Weeder* [11]) traverse a search space of motifs using a well-defined scoring function. Candidate motifs are typically generated and tested during the search, and finally ranked based on either raw score or significance. In the latter category [8,5,7,12], since the motifs are already given, the objective is to find motif instances in input data (if any) by using a scoring function and assessing their significance. Even though the scoring functions and

*This work was carried out during the tenure of an ERCIM fellowship.

significance computations are common components to both categories, they are still fundamentally different. For instance, how to search the motif space is central to the former category, but irrelevant to the second one. On the practical side, methods in the first category usually take scoring functions for which computation can be done relatively easy. This is because during the search, a large number of alternatives need to be evaluated against input data. Methods in the latter category can take relatively complex scoring functions, as motif libraries are relatively small in size, *e.g.* several hundreds for TRANSFAC[4] and about one hundred for JASPAR[1].

Usually scoring functions exploit the following three properties of biological motifs;

(1) Being over-represented (*i.e.*, occurs many times),
(2) Being different from background distribution (*i.e.*, good separation),
(3) Being conserved (*i.e.*, high information content).

Most methods select scalar valued scoring functions that optimize some combination of these properties. Although many functional forms are used, it is unclear which one to choose. Moreover, some methods introduce other parameters into scoring functions, like motif representation language, motif length or motif expectation ratio. However, these are mostly for algorithmic concerns.

Both *de novo* and library methods operate in a single-step manner. Even though some methods (*MEME, MDScan*, etc) have a maximization/refining step as a second step, they are still single-step methods from the point of view of scoring function number. That is, they fix the scoring function beforehand and find and/or rank motifs based on this function. However, selection of scoring function biases the motif discovery as there are many alternatives maximizing different objectives. Though different scoring functions can be combined into a single function [11,10], the same problem remains, *i.e.* how to combine them. The main motivation in this work is using cascades of scoring functions rather than combination.

We propose a two-step paradigm of motif discovery where different scoring functions are used in each step, preferably as independent as possible, thus incorporating various notions of relevance. In this paradigm, the first step is supposed to dramatically reduce the number of possible motif candidates by filtering out non-promising ones. After that, a second step begins and further evaluates promising candidates to decide true motifs by choosing among them. As an example of this paradigm, we present a *de novo* motif discovery method called *TScan*, a word-counting method that is inspired by *MDScan* [10], but that does not require ChIP-chip data.

TScan operates in two steps; 1) ranking all k-mers in the input data set and finding their alignments with respective PWM representations, 2) ranking each PWM with another independent scoring function and computing specificities.

In the first step, each k-mer (oligo) in the input sequences is considered as seed for a motif candidate. For each k-mer, a custom statistical over-representation score is computed. The score takes into account the first two of the three mentioned properties of biological motifs and ranks all k-mers based on that score. The score assigns higher scores to k-mers that are unlikely to belong to background and also occur frequently in the input sequences.

We select high scoring k-mers (*e.g.* top 1%) as motif seeds, and filter out the remaining ones. We compute aligned instances (and respective PWM) for each such k-mer by pooling all exact and partial instances similar to the k-mer (allowing some mismatches). Note that, even though conservation is not explicitly used in the scoring function, it is implicitly included by allowing instances with a few mismatches in alignments.

In the second step, motifs are ranked based on another scoring function. Since the number of motif candidates to be evaluated is relatively small, the scoring function used in this step can be complex and computationally demanding. Besides ranking motifs based on raw scores, they can also be ranked by specificities (p-values) computed from raw scores. A variety of scoring functions can be used in this step. We consider two scoring functions: one custom function based on explicit conservation and over-representation, and another relatively complex function proposed by Frith *et al.* [5] for library based motif discovery.

A similar two-step idea is also exploited in *Weeder*[11], however, the second scoring function is used only to find motif instances not to find motifs. In other words, the second scoring function works exactly as library based methods, *i.e.* finding instances of already identified motifs. Recall that *TScan* uses both functions to decide on motifs.

We have tested *TScan* with real data sets from Harbison *et al.* [3] clusters. The results show that *TScan* is able to find relevant motifs that agree well with the already identified consensuses.

2. Method

Let $G = \{G_g\}_{g=1}^{g=|G|}$ be the set of genes. Define $S = \{S_g : g \in G\}$ as the upstream sequence data such that $S_g =< S_{gl} : l = 1, \ldots, |S_g| >$ where $S_{gl} \in \{A, C, G, T\}$ is the nucleotide in the l'th position and $|S_g|$ is the length of the sequence for g, respectively. Define O_k to be a collection of all k-mers in S, including its reverse complement. Since O_k can contain multiple occurrences of particular oligos, we define the set O_k^u from O_k, where only unique occurrences and one out of multiple occurrences are included. We represent background nucleotide occurrence statistics with a set of discrete distributions, B of Markov order b.

The length k of desired motifs is a user selected parameter with typical values between 6 and 20. Input to step 1 is O_k and output is O_k^q, a subset of top scoring k-mers from O_k^u. Additionally, for each k-mer, aligned instances of k-mers with at most a particular hamming distance H are also output. The input to step 2 is O_k^q with aligned instances of k-mers (and respective PWMs) in it. The output is the n highest scoring (or highest significant) PWMs.

The instances of each of n motifs in input data can be found using aligned instances of respective k-mers or PWMs. In case aligned instances of a k-mer is used, any substring exactly matching in an input sequence to any one in the alignment is considered as a hit. For searching PWMs, however, a similarity threshold must be defined for each PWM or an automated method must be devised to set such a threshold. One such method is given in [12].

2.1. *Step 1*

The probability of obtaining a k-mer $o \in O_k^u$ from B, *i.e.* likelihood, is;

$$P_B(o) = P_B(o(1)) \cdot \prod_{i=2}^{i=k} P_B(o(i)|o(1:i-1)) \tag{1}$$

where $o(i)$ is the nucleotide at the i'th position of o and $o(i:j)$ $(i \leq j)$ is the nucleotide sequence from position i to j, inclusively.

Higher probability of $P_B(o)$ indicates a good conformance of o to the background distribution B. k-mers with lower degree of conformances are good motif candidates. On the other hand, conformance probabilities P_B are usually a weak indication for candidate motifs only by themselves. The over-representation therefore should also be considered. For instance, consider two k-mers, o_1 and o_2 where $P_B(o_1)$ is slightly smaller than $P_B(o_2)$, and suppose that occurrence of o_2 is much larger than that of o_1, then probably o_2 is a better motif candidate than o_1. To account for the over-representation together with the background conformance we compute a score for each k-mer $o \in O_k^u$.

$$Score(o) = \log(\frac{1}{P_B(o)}) \cdot \log(X(o)) \tag{2}$$

where $X(o)$ is the count of k-mer o in the collection O_k.

Intuitively, Formula (2) assigns high scores to k-mers that are over-represented together with having low-conformance to background distribution. Actually, these are the two preconditions for being a motif by definition.

To compute over-representation, simple counting of k-mers can be used. However, simple counting might suffer from being non-distinguishing especially when k is large and input data set is small in size. To alleviate this, we also count k-mers with high similarity to the selected k-mer after weighting. For example, a k-mer with a single mismatch to the selected one can be assumed to be a mutated version of the k-mer. We only consider partial matches within a fixed hamming distance H. Let $\{D^h(o)\}_{h=0}^{h=H}$ be the collection of all k-mers in O_k with the hamming distance h from the k-mer o and denote its cardinality by $|D^h(o)|$. The over-representation of k-mer o is calculated as follows;

$$X(o) = \sum_{h=0}^{h=H} W(h) \cdot |D^h(o)| \tag{3}$$

where $W(h)$ is the weighting factor for the hamming distance h and computed as follows;

$$W(h) = \frac{BinomPdf(k; k, p)}{BinomPdf(k-h; k, p)} \tag{4}$$

where $BinomPdf(k; n, p)$ is the binomial density function with parameters (n, p) evaluated at k. Galas *et al.*[6] employs a similar weighting but they use percentage of matches, rather than probability of random match.

The p in Formula (4) is the probability of a random match of any two nucleotides from the background distribution B. It is simply $B(\cdot) * B(\cdot)^T$, where $B(\cdot)$ is the four dimensional zeroth order row vector of background probabilities and $B(\cdot)^T$ is its transpose. From Formula (4), note that each occurrence of exact match ($h = 0$) is counted as 1 and as h increases respective partial matches are counted with smaller fractions.

We determine the value of H by 3-standard deviations (this roughly corresponds to random partial match with probability 0.0015 or less) from the mean of binomial distribution with parameters (k, p) as follows (as used in *MDScan*);

$$H = k - \lceil kp + 3\sqrt{kp(1-p)} \rceil \tag{5}$$

In Figure 1 the scores (computed with Formula (2)) are shown for four gene clusters from Harbison *et al.*[3] with $k = 8$ and third-order background distribution computed from all of the upstream (at most 500bp) sequences of *Saccharomyces cerevisiae*. The score distributions are almost Normal. Recall that k-mers with score at the right tail of the respective score distributions are good candidates as motif seeds.

We consider a number (*e.g.* 200) or a ratio (*e.g.* 0.5%) of high scoring k-mers to be selected into O_k^q. Since aligned instances of two different seed k-mers can give the same degenerate motif, we identify the strongest one and include only that in O_k^q to avoid redundancy.

2.2. *Step 2*

The objective of this step is to further evaluate aligned instances of all unfiltered k-mers from Step 1 and rank them. The scoring function employed in this step should be as independent as possible from the previous one, but still biologically relevant. To this end we consider two scoring functions: *over-representation/conservation scoring* and *Frith et al. scoring*. The former ranks aligned instances of k-mers based on their cardinality and information content. The latter scores respective PWMs using a relatively complex scoring function. The function takes into account background distribution, conservation and over-representation. The score biases uniform distribution of instances in input data and explicitly takes into account all possible subsets of sequences in input data as TF targets.

2.2.1. *Over-representation/Conservation Scoring*

Given the candidate motif seeds O_k^q, we compute an over-representation/conservation score for all k-mers in this set and rank them based on this score.

For the over-representation/conservation (ORC) score we consider over-representation and conservation (information content) and compute it as follows;

$$ORCScore(o) = IC(O^{h \le H}(o)) \cdot log(|O^{h \le H}(o)|) \tag{6}$$

where $o \in O_k^q$, $O^{h \le H}(o) = \{x : x \in O_k, x \in \{D^h(o)\}_{h=0}^{h=H}\}$, and $IC(O)$ is the information content of alignment O.

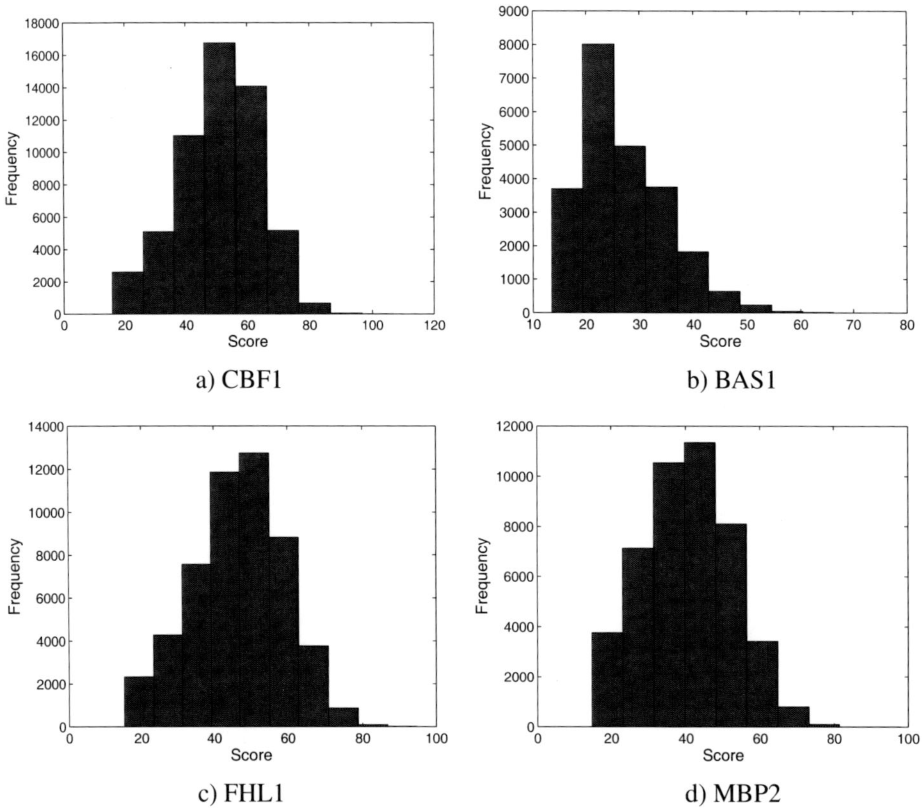

Figure 1. Formula (2) score distribution of all 8-mers of four gene clusters.

Note that Formula (6) favors seed k-mers having high over-representation and high information content (high conservation). Unlike the over-representation computation in Step 1, each occurrence of k-mers in $O^{h \leq H}(o)$ contributes equally with the value of 1, regardless of hamming distance. This is because these occurrences are now treated as aligned binding sites for the respective motif.

All the k-mers in O_k^q are ranked based on their ORC score and highest scoring n k-mers are selected as final seeds for motif generation. Finally, PWMs are computed for each of them from their aligned full/partial matching k-mers.

2.2.2. *Frith et al. Scoring*

Frith *et al.*[5] presented a likelihood based scoring function. It is particularly useful when there is uncertainty of how many sequences contain motif instances for a given motif.

Accumulated likelihood ratio of generating a promoter sequence S_g with a $4 \times k$ PWM matrix M (computed from alignments a k-mer) versus generating it from background distribution B is computed with Formula (7).

$$LR2(M, S_g) = \frac{1}{|S_g^k|} \sum_{o \in S_g^k} \prod_{j=1}^{j=k} \frac{M(o(j), j)}{P_B(o(j))} \quad (7)$$

where S_g^k is the set of all k-mers in S_g (including reverse complements), $P_B(o(j))$ is ratio of nucleotide $o(j)$ in zeroth order background distribution B.

The objective is to get a scalar score for each motif by scoring input set as a whole rather than getting a vector of scores for each motif. Thus, the $LR2$ scores should be combined to give one score. To do this, it is assumed that a motif occurs in an i-subset ($i \leq |S|$) of input promoters, but i is unknown. Assuming it is known and considering all i-subsets equally probable, the following likelihood score can be computed:

$$LR3(M, i) = \frac{1}{\binom{|S|}{i}} \sum_{A} \prod_{S_g \in A} LR2(M, S_g) \quad (8)$$

where A runs over all i combinations. Noting the combinatorial number of i subsets, the authors also provide a method using recurrence relations for computing Formula (8) efficiently. Note that the formula favors uniform distribution of motifs in the considered subset rather than uneven distribution. This is achieved through the product operation of individual scores. Final raw score is computed as an average of $LR3$ scores over all is as given next;

$$LR4(M) = log(\frac{1}{|S|} \sum_{i=1}^{i=|S|} LR3(M, i)) \quad (9)$$

For each seed k-mer in O_k^q we compute raw scores using Formula (9). Motifs can be ranked either by these raw scores or by p-values (specificity). Some different ways of computing p-values is considered in [5]. In case background sequences are known, p-value computation can be carried out empirically by measuring how often we find more extreme $LR4$ scores. In either case we output n motifs.

3. Experiments

Harbison *et al.*[3] identified *S.cerevisiae* target genes for a number of TFs by collecting results from the following resources: ChIP-chip data, published data from literature and phylogenetic conservation. As a result, they defined several dozens of (overlapping) gene clusters for each binding motif. They also confirmed the results by applying several motif discovery programs (*MEME, MDScan*, etc.). In this work we experiment with 7 clusters (Table 1) from their work. We experiment with k-mers where $k = 8$, output 150 candidate motifs from Step 1, and report only $n = 5$ motifs from Step 2. We do not employ repeat maskers.

We first experiment with the *ORCScore* scoring function of Formula (6) used in Step 2. Figure 2 gives the motifs that are most similar to the consensus given in Table 1.[a] We see that results agree well with the consensuses given.

[a] All the motif logos are generated with Weblogo tool located at http://weblogo.berkeley.edu/

Table 1. The clusters used in our experiments.

Regulator	# of genes in the cluster	Consensus motif
CBF1	195	**CACGTG**
FHL1	131	**TGTA[CT]GG[GA]TG**
BAS1	17	**TGACTC**
INO4	32	**CATGTGAA**
MBP1	92	**ACGCGT**
MSN2	74	**[AC]AGGGG[CG]GG**
REB1	99	**CGGGTAA**

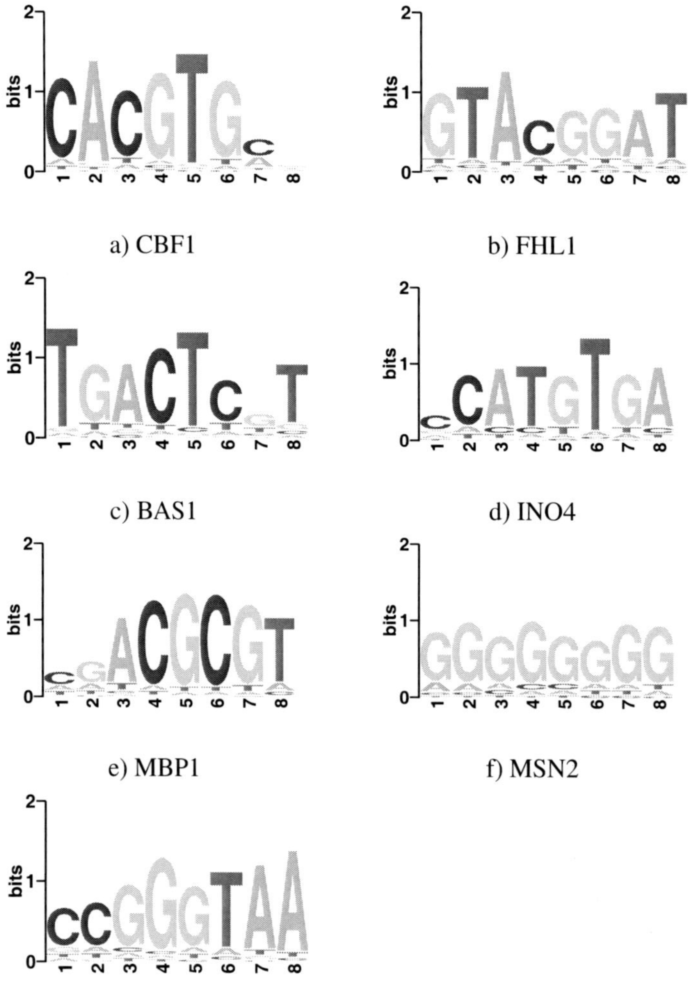

a) CBF1 b) FHL1

c) BAS1 d) INO4

e) MBP1 f) MSN2

g) REB1

Figure 2. Motifs found for 7 clusters when *ORCScore* scoring function was used in Step 2.

We also experiment with the *LR4* scoring function of Formula (9) used in Step 2. Instead of ranking motifs with raw *LR4* scores, we rank motifs with their p-values. p-values are computed empirically using random background as already defined, from 1000 random samples of $S.cerevisiae$ background upstream sequences. In this way, we try to find motifs specific to the respective clusters. Figure 3 gives the motifs that are most similar to the consensus given in Table 1. Except for the *MSN2* cluster, we see that results agree well with the consensuses given. For the *MSN2* cluster we observed that its seed k-mer appears among the 150 motifs output from Step 1, but is still not ranked as one of the top 5 motifs. Thus, it suggests that the respective motif **[AC]AGGGG[CG]GG** is not too specific to *MSN2* cluster.

Even though we have fixed the motif length k to 8 for all clusters, the method is capable of finding motifs having maximum overlap with the differing length consensuses. This shows that the method is not sensitive to consensus length.

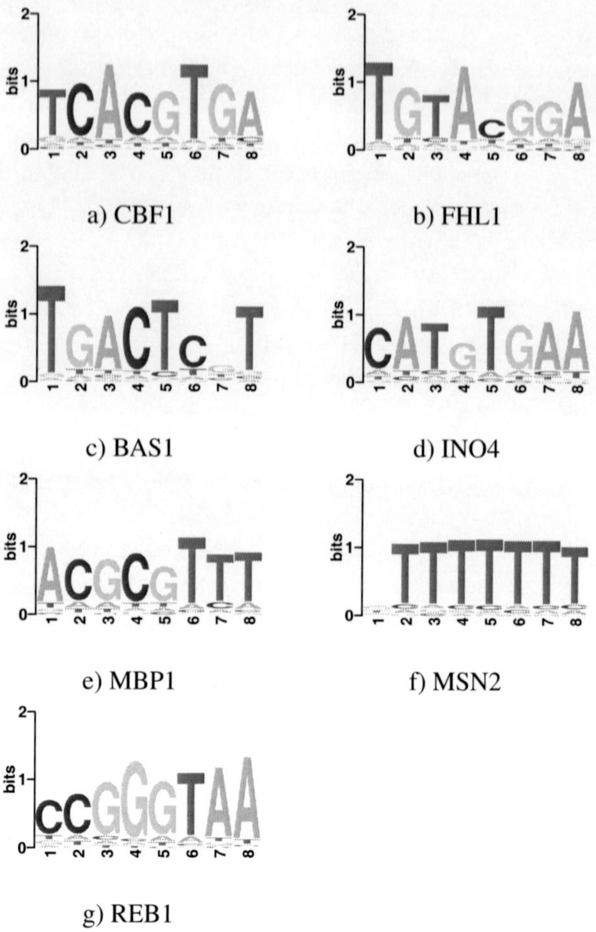

Figure 3. Motifs found for 7 clusters when p-values ranking of *LR4* scoring function was used in Step 2.

In addition to ranking the 150 motifs output from Step 1 with p-values, we have also ranked them based on their raw *LR4* scores. In this case, for the *BAS1* cluster, the highest scoring motifs are not specific, but the general **TATA** box. Actually, this is an expected result from the design of the *LR4* score as it favors motifs appearing almost in all the input sequences. We can reason from this that ranking based on specificity biases towards finding motifs that are specific to input data set, while ranking based on over-representation (like *LR4*) biases towards finding motifs of general regulators. In case the input set contains specific motifs in almost all sequences, both raw *LR4* scores and p-values are expected to give the same results (as we got for *REB1* cluster). Surprisingly, the highest raw scoring motifs for *CBF1*, *FHL1*, *INO4*, *MBP1*, and *MSN2* clusters seem to be junk repetitive sequence with all **T**s, like the one given in Figure 3(f). This suggests employing repeat maskers as a pre-processing step.

4. Conclusion

Scoring functions are key components both while designing *de novo* and library based motif discovery algorithms. Current motif discovery algorithms operate in a single step, meaning they use only one selected scoring function throughout. Therefore, they are quite likely to miss some notions of biological relevance of motifs. To exploit different notions of biological motifs, we have proposed a two-step paradigm for motif discovery. In our approach, different scoring functions are used for each step.

The two steps in our paradigm are executed in order, *i.e.* not interleaved. In this way, the first step serves as data reduction for the second step by removing non-promising candidates. The second step operates on a small subset of motif candidates, thus enabling computationally demanding scoring functions. In this sense, our first step is similar to traditional *de novo* motif discovery algorithms while the second step is more similar to library based methods.

We have proposed a *de novo* motif discovery method, named *TScan*, exploiting the two-step paradigm. It takes the word-counting approach and borrows ideas from *MDScan*. Its objective however is not to replace *MDScan* but to extend it to cases where ChIP-chip data are not available. We have developed a custom scoring function for the first step and proposed two different scoring functions for the second step. The Frith *et al.* scoring function performs particularly well when data are from ChIP-chip experiments, but not necessarily sorted, as it considers all subsets as containing motifs. We would like to note that the scoring functions used in either step can be replaced with other appropriate functions depending on the objective.

The current work has considered both the applicability of the two-step paradigm in general and the proposed application *TScan*. The experimental results are promising. We are currently working on more verification (with a large number of data sets) and comparing the performance of *TScan* against methods that uses a single-step approach.

The two-step paradigm presented here resembles the unsupervised case of the aggregation of weak-classifiers approach used in machine learning. In other words, the different scoring functions may be seen as weak classifiers considering different notions of biological relevance of motifs. Extending our work along this direction, with multiple scoring functions each covering different notions of biological relevance, remains as future work.

References

1. Sandelin A., Alkema W., Engstrom P., Wasserman W.W., and Lenhard B. JASPAR: An open-access database for eukaryotic transcription factor binding profiles. *Nucleic Acids Research*, 32:D91–D94, 2004.
2. Timothy L. Bailey and Charles Elkan. Unsupervised Learning of Multiple Motifs in Biopolymers using Expectation Maximization. *Machine Learning*, (21):51–80, 1995.
3. Harbison CT, Gordon DB, Lee TI, Rinaldi NJ, Macisaac KD, Danford TW, Hannett NM, Tagne JB, Reynolds DB, Yoo J, Jennings EG, Zeitlinger J, Pokholok DK, Kellis M, Rolfe PA, Takusagawa KT, Lander ES, Gifford DK, Fraenkel E, and Young RA. Transcriptional regulatory code of a eukaryotic genome. *Nature*, 431(7004):99–104, 2004.
4. Windenger E., Chen X., Hehl R., Karas H., Liebich I., Matys V., Meinhardt T., Pruss M., Reuter I., and Schacherer F. JASPAR: An open-access database for eukaryotic transcription factor binding profiles. *Nucleic Acids Research*, 28:316–319, 2000.
5. Martin C. Frith, Yutao Fu, Liqun Yu, Jiang-Fan Chen, Ulla Hansen, and Zhiping Weng. Detection of functional DNA motifs via statistical over-representation. *Nucleic Acids Research*, 32(4):1372–1381, 2004.
6. David J. Galas, Mark Eggert, and Michael S. Waterman. Rigorous Pattern-recognition Methods for DNA Sequences. *J. Mol. Biol.*, 186:117–128, 1985.
7. Jason D. Hughes, Preston W. Estep, Saeed Tavazoie, and George M. Church. Computational Identification of Cis-regulatory Elements Associated with Groups of Functionally Related Genes in Saccharomyces Cerevisiae. *Journal of Molecular Biology*, (296):1205–1214, 2000.
8. A.E. Kel, E. Gobling, I. Reuter, E. Cheremushkin, O.V. Kel-Margoulis, and E. Windenger. MATCH: A tool for searching transcription factor binding sites in DNA sequences. *Nucleic Acids Research*, 31(13):3576–3579, 2003.
9. X. Liu, D.L. Brutlag, and J.S. Liu. Bioprospector: Discovering Conserved DNA Motifs in Upstream Regulatory Regions of Co-expressed Genes. In *Proc. of Pacific Symposium on Biocomputing*, 2001.
10. X. Shirley Liu, Douglas L. Brutlag, and Jun S. Liu. An Algorithm for Finding Protein-DNA Binding Sites with Applications to Chromatin-Immunoprecipitation Microarray Experiments. *Nature Biotechnology*, 20:835–839, 2002.
11. Giulio Pavesi, Paolo Mereghetti, Giancarlo Mauri, and Graziano Pesole. Weeder Web: discovery of transcription factor binding sites in a set of sequences from co-regulated genes. *Nucleic Acids Research*, 32:W199–W203, 2004.
12. Elkon R., Linhart C., Sharan R., Shamir R., and Shiloh Y. Genome-wide in silico identification of transcriptional regulators controlling the cell cycle in human cells. *Genome Research*, 13:773–780, 2003.

REDUNDANCY ELIMINATION IN MOTIF DISCOVERY ALGORITHMS*

HENRY LEUNG

Department of Computer Science, The University of Hong Kong, Pokfulam, Hong Kong, China

FRANCIS CHIN

Department of Computer Science, The University of Hong Kong, Pokfulam, Hong Kong, China

Abstract: The problem of finding motifs in binding sites is very important to the understanding of gene regulatory networks. However, when predicting a set of motifs, existing algorithms suffer the problem of either predicting many redundant motifs (motifs with similar binding sites) or, at the other extreme, missing the hidden motif. In this paper, we formulate the Motif Redundancy Problem (MRP) to model this kind of problem and introduce an algorithm called RME (Redundancy Motif Elimination) for solving MRP. Experimental results on real biological data show that a standard EM-based motif discovery algorithm enhanced with RME has a better performance than the popular motif discovery algorithm MEME.

1. Introduction

A *gene* is a segment of DNA that is the blueprint for protein. In most cases, genes do not work alone; rather, they cooperate to produce different proteins for a particular function. In order to start the protein decoding process (*gene expression*), a molecule called *transcription factor* will bind to a short region (*binding site*) preceding the gene. One transcription factor can bind to the binding sites of several genes to cause these genes to co-express. These binding sites have similar patterns called *motifs*. Finding motifs and the binding sites from a set of DNA sequences, which represent the *promoter regions* of co-expressed genes, is a critical step for understanding the *gene regulatory network*.

In order to discover motifs, we must first have a model to represent the motif. There are two popular models: string representation [3,5-7,11,12,15,17,19,20,22-28] and matrix representation [1,2,8,9,13,14,16,18]. String representation is the most basic representation which uses a length-l string of symbols (or nucleotides) 'A', 'C', 'G' and 'T' to describe a motif. To improve the representation's descriptive power, wildcard symbols [5,22,26] can be introduced into the string to represent choices from a subset of symbols at a particular position (e.g. 'K' can denote 'G' or 'T'). Matrix representation further improves descriptive power. In the matrix model, motifs of length l are represented by *position weight matrices* (PWMs) or *position specific scoring matrices* (PSSMs) of size $4 \times l$ with the jth column of the matrix, which has four elements corresponding to the four

* The research was supported in parts by the RGC grant HKU 7120/06E.

nucleotides, effectively giving the occurrence probability of each of the four nucleotides at position j.

When discovering motif in matrix representation, researchers usually assume the motif matrix with the largest likelihood, calculated based on some probability model [1,13,14,16], is the hidden motif. However, motif discovery algorithms usually output a set of predicted motifs instead of a single motif with the highest likelihood because:

1. The input DNA sequences may contain binding sites of several transcription factors. Therefore, the algorithms should discover a motif for each of these transcription factors.
2. The motif matrix with the highest likelihood may not be a meaningful motif to the biologist and is considered only to be an over-represented pattern that occurs accidentally in the input sequences (noise), or alternatively, occurs in every part of the whole genome.
3. Biologists need to perform experiments to verify predicted motifs. Sometimes one experiment can be performed on several predicted motifs simultaneously in order to verify the correct one.

When motif discovery algorithms predict a set of motifs, they generally suffer the problem that some of the predicted motifs are very similar (*redundant motifs*) in the sense that they represent almost the same set of binding sites. For example, the motif matrices (predicted for AS-CT3) shown in Table 1, with M_1 being the hidden motif, are very similar:

Table 1. Example of similar motifs.

Motif	Log Likelihood	PSSM	Binding sites pattern
M_1	–9833.12	$\begin{matrix}A\\C\\G\\T\end{matrix}\begin{pmatrix}0.21 & 0.49 & 0.19 & 0.14 & 0.12 & 0.17\\0.51 & 0.27 & 0.13 & 0.20 & 0.19 & 0.19\\0.16 & 0.07 & 0.23 & 0.48 & 0.47 & 0.19\\0.11 & 0.17 & 0.44 & 0.16 & 0.22 & 0.45\end{pmatrix}$	CAGGTG [seq 1:355-360] CAGGTG [seq 2:336-341]
M_2	–9833.17	$\begin{matrix}A\\C\\G\\T\end{matrix}\begin{pmatrix}0.48 & 0.23 & 0.19 & 0.17 & 0.14 & 0.18\\0.25 & 0.47 & 0.13 & 0.21 & 0.22 & 0.49\\0.14 & 0.14 & 0.49 & 0.53 & 0.25 & 0.21\\0.13 & 0.16 & 0.18 & 0.08 & 0.39 & 0.13\end{pmatrix}$	CAGGTG [seq 1:355-360] CAGGTG [seq 2:336-341] CAGGGG [seq 2:314-319]
M_3	–9832.36	$\begin{matrix}A\\C\\G\\T\end{matrix}\begin{pmatrix}0.65 & 0.00 & 0.00 & 0.06 & 0.00 & 0.01\\0.03 & 0.00 & 0.13 & 0.00 & 0.01 & 0.02\\0.27 & 0.89 & 0.79 & 0.00 & 0.82 & 0.21\\0.04 & 0.10 & 0.08 & 0.94 & 0.17 & 0.76\end{pmatrix}$	AGGTGT [seq 1:356-361] AGGTGG [seq 2:337-341] AGGTGT [seq 2:31-36]

M_1 and M_2 have two binding sites in common. Although M_3 is different from M_1 and M_2, its first two binding sites overlap with the first two binding sites of M_1 and M_2 with a one base pair shift. Although these three motifs have high likelihoods, motif discovery algorithms should not output them all because they represent almost the same set of binding sites and would increase the size of output unnecessarily without meaningful benefit.

Some motif discovery algorithms [16] reduce the number of redundant motifs in the output by replacing a set of redundant motifs by a motif in that set. Other algorithms [14] (e.g. MEME) solve this problem by finding motifs one by one and by making, at each iteration, adjustments so as to reduce the probability that a binding site of an already-discovered motif is considered again. While this approach will help to reduce redundant motifs, accuracy may also suffer because the hidden motif might not be discovered if its binding sites happened to overlap with the binding sites of some previously discovered motifs. This approach depends very much on the order of motifs being discovered. For example, consider the transcription factor "CF2-1" of fruit fly and the following discovered motifs arranged in the order of decreasing log-likelihood:

1	TTTTTTTT	log likelihood = –6399.45
2	GCGCCTGC	log likelihood = –6400.02
3	GCCCCCGC	log likelihood = –6403.37
	…	
19	GTTTTATT	log likelihood = –6409.44

The correct motif (ranked 19) will never be discovered in MEME if motif 1 is discovered first, nor, in those algorithms where redundant motifs are eliminated, because of its overlap with the first motif.

Moreover, the simple approach in which we allow for redundancy but limit the size of the output may also not work in situations such as that of the transcription factor "bcd" of fruit fly (see below) where the correct motif (ranked 281) is ranked far down the list.

1	CCCAACCC	log likelihood = –33744.8
2	CCAATCCC	log likelihood = –33753.4
3	CCCGATCC	log likelihood = –33755.8
	…	
281	TGGATTAG	log likelihood = –33870

In this paper, we introduce a novel way to select the "best" motifs among all possible motifs (redundant or otherwise) based on their likelihood of being the hidden motif and also their pair-wise redundancies. We first formulate the *Motif Redundancy Problem* (MRP) as the problem of picking a set of motifs for output such that accuracy will not be affected too much. However, since MRP is NP-hard, we cannot find the optimal solution of MRP in polynomial time unless P equals NP. Thus, we introduce a heuristic algorithm RME (Redundant Motif Elimination) to solve MRP. We show the usefulness of RME by comparing the performance of the popular software MEME against a simple EM algorithm enhanced by using RME to eliminate redundant motifs. We find that a simple EM algorithm can outperform MEME with the help of RME. Moreover, RME does successfully output the correct motif for the cases described above of "CF2-1" and "bcd".

This paper is organized as follows. In Section 2, we briefly describe how to calculate the likelihood of a matrix being the hidden motif. In Section 3, we introduce MRP. The heuristic algorithm RME for solving MRP is described in Section 4. Experimental results on real biological data comparing are given in Section 5, followed by concluding remarks in Section 6.

2. Maximizing Likelihood

Existing algorithms using PSSM matrix representation discover the motif matrix with the maximum likelihood of being the hidden motif based on the finite mixture model [1]. Conceptually, they break up the input sequences into length-l (overlapping) substrings. For example, a length-n sequence can be broken up into $w = n - l + 1$ length-l substrings. Let $X = (X_1, X_2, \ldots, X_w)$ be all length-l substrings in the input where each substring can occur more than once in X if the same pattern appears in more than one position. The finite mixture model assumes that each substring in X belongs to either a background (non-motif) substring or an instance of the hidden motif. The prior probability that a substring belongs to the background substrings (generated according to the background probability) is λ_b and the prior probability that a substring belongs to binding sites (generated according to the hidden matrix) is $1 - \lambda_b$.

Let $Z = (Z_1, Z_2, \ldots, Z_w)$ be the missing data that determines whether X_i is generated according to the background probability B or the hidden matrix M.

$$Z_i = \begin{cases} 1 & X_i \text{ is generated according to } B \\ 0 & X_i \text{ is generated according to } M \end{cases}$$

The conditional probability that a substring X_i is generated according to the background probability $B = (b(\text{A}), b(\text{C}), b(\text{G}), b(\text{T}))$ is

$$p(X_i \mid Z_i = 1, B) = \prod_{j=1}^{l} b(X_i[j])$$

where $X_i[j]$ is the jth nucleotide in X_i.

The conditional probability that a substring X_i is generated according to a $4 \times l$ hidden matrix M is

$$p(X_i \mid Z_i = 0, M) = \prod_{j=1}^{l} M(X_i[j], j)$$

where $M(\alpha, j)$ is the probability that nucleotide α occurs at the jth position of an instance of M. Note that $\sum_\alpha M(\alpha, j) = 1$.

By the assumption of independent substrings X, the joint conditional density of X and Z being generated according to the finite mixture model with parameters B, M and λ_b is

$$\begin{aligned} & p(X, Z \mid B, M, \lambda_b) \\ & = \prod_{i=1}^{w} \left(p(X_i, Z_i \mid B, M, \lambda_b) \right) \\ & = \prod_{i=1}^{w} \left(p(X_i \mid Z_i, B, M, \lambda_b) p(Z_i \mid B, M, \lambda_b) \right) \\ & = \prod_{i=1}^{w} \left(\left[\lambda_b \prod_{j=1}^{l} b(X_i[j]) \right]^{Z_i} \left[(1 - \lambda_b) \prod_{j=1}^{l} M(X_i[j], j) \right]^{1 - Z_i} \right) \end{aligned}$$

The likelihood of a particular B, M, λ_b being the hidden parameters of the finite mixture model given the joint distribution of the substring X and the missing data Z is defined as

$$L(B,M,\lambda_b \mid X,Z) = p(X,Z \mid B,M,\lambda_b) \tag{1}$$

So, the log of the likelihood, or *log likelihood*, is therefore

$$\log L(B,M,\lambda_b \mid X,Z) = \sum_{i=1}^{w}\left\{Z_i\left[\log(\lambda_b)+\sum_{j=1}^{l}\log(b(X_i[j]))\right]+(1-Z_i)\left[\log(1-\lambda_b)+\sum_{j=1}^{l}\log(M(X_i[j],j))\right]\right\} \tag{2}$$

The goal of existing algorithms using matrix representation is to discover the B, M, λ_b with the maximum likelihood (or log likelihood).

3. The Motif Redundancy Problem

In practice, biologists want to get a set of predicted motif matrices with high likelihood instead of a single matrix with the highest likelihood (as discussed in the introduction). However, existing motif discovery algorithms either allow the output of many redundant motifs (i.e. motifs representing almost the same set of binding sites), which makes the output size unnecessarily large, or try to eliminate redundant motifs and in the process risk also eliminating the hidden motif. We introduce the *Motif Redundancy Problem* (MRP) to reduce the size of the output with the least reduction in accuracy.

3.1. *The motif redundancy problem*

Assume the positions of the planted binding sites of the hidden matrix M^* are known. The accuracy of a predicted motif with matrix M is measured by its *score* $s(M,M^*)$ expressed as follows [3]:

$$s(M,M^*) = \frac{|\text{sites of } M \cap \text{ sites of } M^*|}{|\text{sites of } M \cup \text{ sites of } M^*|} \tag{3}$$

We say a planted binding site at position $[x, x + l - 1]$ is *correctly predicted* if that planted binding site overlaps with at least one predicted site $[y, y + l - 1]$, i.e. $[x, x + l - 1] \cap [y, y + l - 1]$ is non-empty. The score $s(M,M^*)$ is in the range of [0,1]. When all the planted binding sites are correctly predicted without any mis-prediction, score = 1. When no planted binding site is predicted correctly, score = 0. Since biologists will select the best motif as the hidden motif, the *accuracy of a motif discovery algorithm* can be measured by the maximum score obtained from its set of predicted motifs, i.e. $\max\{s(M,M^*) \mid M \in S\}$ where S is the set of motif matrices predicted by the algorithm. Note that accuracy increases with the size of output of any motif discovery algorithm.

3.2. *Formal definition*

Given a set S of motif matrices M_i (where i = 1, …, q) and their binding sites and corresponding likelihood $L_i = L(B,M_i,\lambda_b \mid X,Z)$. Assuming the hidden motif is one of the q matrices in S, we can estimate the probability $P(M_j \mid X)$ that M_j is the hidden matrix of the data set X by its likelihood $L_i = L(B, M_i, \lambda_b \mid X, Z)$.

Since $L_i = L(B, M_i, \lambda_b \mid X, Z) = p(X, Z \mid B, M_i, \lambda_b) \propto P(M_i \mid X)$, we have

$$P(M_j \mid X) = \frac{L_j}{\sum_{k=1}^{q} L_k} \tag{4}$$

As the score of M_i is $s(M_i, M_j)$ if M_j is the hidden matrix, the *expected score* of M_i is

$$E(M_i) = \sum_{j=1}^{q} s(M_i, M_j) P(M_j \mid X)$$

Given a set S of motif matrices, their binding sites and likelihoods, $E(S)$ can be calculated with Eq (1), (3) and (4). That is,

$$E(S) = \sum_{j=1}^{q} \max_{M \in S} \{ s(M, M_j) \} P(M_j \mid X) \tag{5}$$

The *Motif Redundancy Problem* (MRP) is defined as follows:

Given a set of input sequences, a set of motif matrices M_i (where i = 1, …, q), their binding sites and corresponding likelihood $L_i = L(B, M_i, \lambda_b \mid X, Z)$, find a subset $S \subseteq \{M_i, i = 1, \ldots, q\}, |S| = m$ such that $E(S)$ is maximized.

Note that a set of redundant motifs usually has a lower score on average, because $s(M, M_j)$ of each redundant motif M is almost the same and there will be many motifs M_j not in S with a low $\max\{s(M, M_j) \mid M \in S\}$ value. This means there is a higher chance of missing the hidden motif. Therefore, a set S with the largest expected score $E(S)$ tends to contain non-redundant motifs.

4. Algorithm

We can show that MRP is a NP-hard problem by transforming the Set Covering Problem to MRP (shown in the Appendix). Therefore, it is not possible to find a polynomial time algorithm to solve MRP unless P equals NP. We apply a heuristic algorithm RME (which stands for *Redundancy Motif Elimination*) to solve MRP. RME finds the subset S of motifs with large expected score $E(S)$ by selecting motifs that give the largest increase in $E(S)$ one by one until the size of S is m. This is essentially a greedy approach. Although RME is simple, experimental results (see Section 5) on real biological data show that it works well in practice. Algorithm RME is shown in below.

Algorithm RME

1. $S = \{\arg\max_{M_i} E(M_i)\}, i = 1, \ldots, q$
2. For i = 2 to m
3. $S = \{\arg\max_{M_j} E(S \cup M_j)\}, j = 1, \ldots, q$
4. output S

First we begin with S containing the motif matrix with the highest expected score $E(M_i)$. At each step, we add a new motif matrix to S such that the new expected score $E(S)$ has the largest increase in value.

In order to illustrate how RME works, let us consider the following example.

Example. Suppose $\{M_1, M_2, M_3\}$ is a set of three predicted matrices with 0.45, 0.35 and 0.2 as their corresponding $P(M_j \mid X)$ where M_1 and M_2 are two redundant motifs. Thus, $s(M_1, M_2) = s(M_2, M_1) = 0.8$ and $s(M_3, M_1) = s(M_3, M_2) = s(M_1, M_3) = s(M_2, M_3) = 0.2$. RME would first choose M_1 since $E(M_1) = 1 * 0.45 + 0.8 * 0.35 + 0.2 * 0.2 = 0.77$ (by Eq (5)) has the highest expected score. If $m = 2$, the second motif matrix will be M_3 instead of M_2 even though M_2 has a higher likelihood since $E(\{M_1, M_2\}) = 1 * 0.45 + 1 * 0.35 + 0.2 * 0.2 = 0.84$ and $E(\{M_1, M_3\}) = 1 * 0.45 + 0.8 * 0.35 + 1 * 0.2 = 0.93$.

5. Experimental Results

We have implemented the standard EM algorithm [14] for discovering motifs and RME on C++ and have performed experiments on real biological data for fruit fly (Drosophila) and yeast from the database TRANSFAC (http://www.gene-regulation.com) and SCPD (http://rulai.cshl.edu/SCPD/). For each transcription factor, we searched for all genes regulated by that transcription factor and used the 450 base pairs (bp) upstream and 50 bp downstream from the transcriptional start site of these genes as the input sequences.

MEME [1], a popular motif discovery program (which is based on a more complicated EM-algorithm), was compared to the performance of a standard EM algorithm enhanced with RME and the standard EM algorithm without RME. Each of the three algorithms predicted 30 motifs with length equal to the published motif of the corresponding transcription factor. The standard EM algorithm first generated 300 predicted motifs and, with and without RME, 30 motifs were picked as output. The performance of these algorithms on each transcription factor was measured by the following formula for accuracy and is shown in Tables 2 (fruit fly) and Table 3 (yeast) below.

$$\text{accuracy} = \frac{|\text{predicted sites} \cap \text{published sites}|}{|\text{predicted sites} \cup \text{published sites}|}$$

In the 47 experiments of the fruit fly, all three algorithms failed to predict any published binding site correctly in 4 data sets (which are not shown in Table 2). For the remaining 43 data sets, the standard EM algorithm with RME has better performance (i.e. higher score) in 28 data sets and equal performance (i.e. the same score) in 4 data sets, while MEME had a better performance in 10 data sets and the standard EM algorithm without RME has better performance in 1 data set only. Moreover, the average score of the standard EM algorithm with RME is 0.2787, which is higher than the average score of MEME of 0.1934 and the average score of the standard EM algorithm without RME of 0.0873.

Table 2. Experimental results on real biological data for transcription factors of fruit fly (Drosophila) in TRANSFAC.

Factor	l	MEME	EM	RME	Factor	l	MEME	EM	RME
Ac	8	**1**	0.0053	0.4	Exd	20	0	0	**0.3333**
adf-1	11	**0.1667**	0.0769	0.15	Ftz	12	0.1724	0.1351	**0.2424**
antp	7	0	0.0476	**0.2**	FTZ-F1	7	0.3333	0	**0.6667**
AP-1	9	0	0	**0.1429**	GAGA	11	0.0870	0.1	**0.2222**
AS-CT3	6	**1**	0.0435	**1**	GCM	13	0.25	0.4286	**0.5**
bcd	8	0.2941	0.0130	**0.4375**	H	10	0	0.1667	**0.25**
BEAF-32B	5	0	**0.7143**	0.25	Hb	10	0.1429	0.1875	**0.2941**
Bfactor	4	**0.1**	0	0	HSTF	15	**0.1667**	0	**0.1667**
CF1	9	0	0	**0.2**	Kr	10	**0.3077**	0.0526	0.2778
CF2-I	8	0	0	**0.1667**	Sc	8	**0.5**	0.0526	0.4
Ci	9	0.2857	0.0909	**0.3636**	Sn	13	**0.2857**	0	0.2105
Cut	7	0	0	**0.1667**	Su_Hw	12	0	0.1	**0.1**
D_MEF2	10	0	0	**0.1667**	TAB	15	**0.3333**	0.1111	0.1333
Dl	11	0.0870	0.0909	**0.1379**	TBP	7	0.25	0	**0.4**
Da	6	0	0	**0.3333**	TII	8	0.0769	0.0938	**0.1304**
DREF	14	**0.1429**	0	0	Ttk69k	8	**0.3333**	0.2222	**0.3333**
dri	10	0	0	**0.1429**	Ubx_a	19	0	0.0833	**0.25**
DTF-1	6	0.25	0.1111	**0.75**	Zen-1	8	0.1	0.1538	**0.1579**
E74A	17	**0.4**	0.1143	0.2941	Zen-2	8	0.1111	0.1111	**0.25**
EcR	7	**0.3333**	0	**0.3333**	Zeste	11	0.1026	0.1290	**0.3191**
Elf-1	8	**0.5**	0.1	0.1429	Zeste_b	11	0.1026	0.1290	**0.3191**
En	7	0.1	0.0909	**0.25**					
Average score							0.1934	0.0873	**0.2787**
Number of times getting the best performance							14	1	**32**

Table 3. Experimental results on real biological data for transcription factors of yeast in SCPD.

Factor	l	MEME	EM	RME	Factor	l	MEME	EM	RME
13nt	13	**0.75**	0.375	**0.75**	GFI	13	0.2	0	**0.2222**
ABF1	13	0.0870	0.0082	**0.125**	HAP1	12	**0.5833**	0	0.5625
ACE2	6	0.3333	0.3	**0.6**	HAP2	7	**0.3333**	0.0833	**0.3333**
ADR1	5	**0.6667**	0.3333	**0.6667**	HSE_2	8	**0.7**	0	0.6
AP1	7	**1**	0	0.25	IRE	32	**1**	0.1053	0.5
BAS1	7	0.3889	0	**0.4444**	LEU	10	0.6667	0.3333	**1**
BAS2	6	0.25	0	**0.3333**	MAT2	9	0.3333	0.0690	**0.4375**
CCBF	7	**0.9375**	**0.9375**	**0.9375**	MCM1	5	**0.4590**	0	0
CPF1	7	**0.6667**	0.1818	0.6	MIG1	12	**0.3333**	0.0164	0.2174
CSRE	12	0.3333	0.0731	**0.3636**	NBF	9	0.5	0.375	**0.5714**
CURE	7	**0.6667**	0.5714	**0.6667**	SFF	10	**0.2**	0	0.1875
GAL4	17	**0.8666**	0.8235	0.8235	SWI5	6	0.4444	0	**0.5714**
GATA	6	0.3913	**0.45**	**0.45**	UASCAR	11	0.25	0	**0.5**
GCFAR	6	0.5714	0.5714	**0.8**	UASGABA	19	0.4	**0.6667**	**0.6667**
GCN4	6	0.3333	0.0222	**0.3571**	UESPHR	9	**0.6667**	0	0.4285
GCR1	5	**0.1818**	0	0.0526					
Average score							**0.4998**	0.2030	0.4845
Number of times getting the best performance							16	3	**21**

In the experiments, the performance of the standard EM algorithm was much improved with the application of RME (except BEAF-32B). Instead of selecting 30 redundant motifs with the highest score generated by the EM algorithm, RME selected 30 non-redundant motifs, which increased the probability of predicting the hidden motifs. Although the standard EM algorithm has a worse performance than the sophisticated MEME in most cases, it outperformed MEME after applying RME.

In the 32 experiments of yeast, all three algorithms failed to predict any published binding site correctly in 1 data set (which is not shown in Table 3). For the remaining 31 data sets, the standard EM algorithm with RME had better performance (higher score) in 15 data sets and the same score in 6 data sets, while MEME had better performance in 10 data sets. The average score of the standard EM algorithm with RME was 0.4845, which was slightly lower than the average score of MEME of 0.4995 but much higher than the average score of the standard EM algorithm of 0.2030.

As shown by the results given in Tables 2 and 3, the standard EM algorithm without RME had a worse performance than MEME in all cases (except GATA and UASGABA). However, the standard EM algorithm had a similar performance to MEME after applying RME. This indicates that algorithm RME can indeed improve the performance of motif discovery algorithms.

6. Concluding Remarks

Many motif discovery algorithms use heuristics to search for motif matrices according to some performance criteria. In most cases, many redundant (similar) motif matrices will be found based on local optimal values. We introduce the motif redundancy problem (MRP) so as to find the best fixed-size output with highest accuracy. Even though MRP is NP-complete, we have demonstrated in this paper that a simple greedy approach (RME) can already bring an improvement in the accuracy of the output. However, this is only a preliminary result, and there are at least two other approaches one could take to make further improvements: (1) improving the heuristic used in solving MRP; and (2) the elimination of the assumption that the hidden motif is among one of the predicted motif. We would include these improvements in our future paper.

Appendix

In this section, we will show how to reduce the Set Covering Problem (SC) to the Motif Redundancy Problem (MRP). The Set Covering Problem is defined as follows:

Given a finite set $X = \{x_1, \ldots, x_n\}$ and a family F of subsets $X_1, \ldots, X_p$ such that every elements x_i in X belongs to at least one subsets in the family. We want to determine whether there is a subset $C \subseteq F, |C| = k$ such that $X = \cup_{X_i \in C} X_i$.

We reduce SC to MRP in the following manner. Let $W = p + n$. We construct $W + pW$ motifs. The first p motifs (set P) represent the p subsets $X_1, \ldots, X_p$, the next n motifs (set N) represent the n elements $x_1, \ldots, x_n$ in set X and the last pW are dummy motifs (set D).

Each motif has equal likelihood $1/(W + pW)$ and has exactly $W + pW$ binding sites. There are two ways in which a motif can have exactly one binding sites overlapped with another motif:

1. The motif representing X_i has one binding site overlapping with the motif representing x_j if and only if $x_j \in X_i$.
2. The motif representing X_i has one binding site overlapping with W distinct dummy motifs.

Except as a result of applying the above two rules, no other binding-site overlaps are allowed.

In particular, the $W + pW$ binding sites for each motif are constructed in the following manner. Each binding site is of length-(W+pW) DNA sequence with nucleotide 'T' at every position except the yth position in the case of yth motif and the zth position if the binding site of this motif is meant to overlap with that of the yth motif (according to one of the above two rules). Nucleotide 'A' appears at such (yth and zth) positions instead.

SC can be reduced to MRP by finding a subset S of motifs, $|S| = k$ such that $E(S)$ is maximized. There is a solution for SC if and only if

$$E(S) = \frac{1}{W + pW}\left(k + (n + kW)\frac{1}{2(W + pW) - 1}\right)$$

References

1. T. Bailey and C. Elkan. Unsupervised learning of multiple motifs in biopolymers using expectation maximization. *Machine Learning*, 21:51–80, 1995.
2. Y. Barash, G. Bejerano and N. Friedman. A simple hyper-geometric approach for discovering putative transcription factor binding sites. *WABI*, 278–293, 2001.
3. J. Buhler and M. Tompa. Finding motifs using random projections. *RECOMB*, 69–76, 2001.
4. M.L. Bulyk, P.L.F. Johnson and G.M. Church. Nucleotides of transcription factor binding sites exert interdependent effects on the binding affinities of transcription factors. *Nuc. Acids Res.*, 30:1255–1261, 2002.
5. F. Chin and H. Leung. An Efficient Algorithm for String Motif Discovery. *APBC*, 79–88, 2006.
6. F. Chin and H. Leung. An Efficient Algorithm for the Extended (*l*,*d*)-Motif Problem With Unknown Number of Binding Sites. *BIBE*, 11–18, 2005.
7. F. Chin and H. Leung. Voting Algorithms for Discovering Long Motifs. *APBC*, 261–271, 2005.
8. F. Chin, H. Leung, S.M. Yiu, T.W. Lam, R. Rosenfeld, W.W. Tsang, D. Smith and Y. Jiang. Finding Motifs for Insufficient Number of Sequences with Strong Binding to Transcription Factor. *RECOMB*, 125–132, 2004.
9. G.Z. Hertz and G.D. Stormo. Identification of consensus patterns in unaligned dna and protein sequences: a large-deviation statistical basis for penalizing gaps. *The*

Third International Conference on Bioinformatics and Genome Research, 201–216, 1995.

10. J.D. Hughes, P.W. Estep, S. Tavazoie and G.M. Church. Computational identification of cis-regulatory elements associated with groups of functionally related genes in Saccharomyces cerevisiae. *Journal of Molecular Biology*, 296(5):1205–14, 2000.
11. U. Keich and P. Pevzner. Finding motifs in the twilight zone. *RECOMB*, 195–204, 2002.
12. S. Kielbasa, J. Korbel, D. Beule, J. Schuchhardt and H. Herzel. Combining frequency and positional information to predict transcription factor binding sites. *Bioinformatics*, 17:1019–1026, 2001.
13. C. Lawrence, S. Altschul, M. Boguski, J. Liu, A. Neuwald and J.Wootton. Detecting subtule sequence signals: a Gibbs sampling strategy for multiple alignment. *Science*, 262:208–214, 1993.
14. C. Lawrence and A. Reilly. An expectation maximization (em) algorithm for the identification and characterization of common sites in unaligned biopolymer sequences. *Proteins: Structure, Function and Genetics*, 7:41–51, 1990.
15. H. Leung and F. Chin. Algorithms for Challenging motif problems. *JBCB*, 43–58, 2005.
16. H. Leung and F. Chin. Finding Exact Optimal Motif in Matrix Representation by Partitioning. *Bioinformatics*, 22:ii86–ii92, 2005.
17. H. Leung and F. Chin. Generalized Planted (l,d)-Motif Problem with Negative Set. *WABI*, 264–275, 2005.
18. H. Leung, F. Chin, S.M. Yiu, R. Rosenfeld and W.W. Tsang. Finding Motifs with Insufficient Number of Strong Binding Sites. *Jour. Comp. Biol.*, 12(6):686–701, 2005.
19. M. Li, B. Ma and L. Wang. Finding similar regions in many strings. *Journal of Computer and System Sciences*, 65:73–96, 2002.
20. S. Liang. cWINNOWER Algorithm for Finding Fuzzy DNA Motifs. *Computer Society Bioinformatics Conference*, 260–265, 2003.
21. T.K. Man and G.D. Stormo. Non-independence of Mnt repressor-operator interaction determined by a new quantitative multiple fluorescence relative affinity (QuMFRA) assay. *Nuc. Acids Res.*, 29:2471–2478, 2001.
22. G. Pesole, N. Prunella, S. Liuni, M. Attimonelli and C. Saccone. Wordup: an efficient algorithm for discovering statistically significant patterns in dna sequences. *Nucl. Acids Res.*, 20(11):2871–2875, 1992.
23. P. Pevzner and S.H. Sze. Combinatorial approaches to finding subtle signals in dna sequences. *The Eighth International Conference on Intelligent Systems for Molecular Biology*, 269–278, 2000.
24. S. Rajasekaran, S. Balla and C.H. Huang. Exact algorithms for planted motif challenge problem. *APBC*, 249–259, 2005.
25. S. Sinha. Discriminative motifs. *The Sixth Annual International Conference on Computational Biology*, 291–298, 2002.
26. S. Sinha and M. Tompa. A statistical method for finding transcription factor binding sites. *The Eighth International Conference on Intelligent Systems for Molecular Biology*, 344–354, 2000.

27. K.T. Takusagawa and D.K. Gifford. Negative information for motif discovery. *PSB*, 360–371, 2004.
28. M. Tompa. An exact method for finding short motifs in sequences with application to the ribosome binding site problem. *The Seventh International Conference on Intelligent Systems for Molecular Biology*, 262–271, 1999.
29. Y. Xing, J.D. Fikes and L. Guarente. Mutations in yeast HAP2 HAP3 define a hybrid CCAAT box binding domain. *EMBO Journal*, 12:4647–4655, 1993.
30. X. Zhao, H. Huang and T.P. Speed. Finding Short DNA Motifs Using Permuted Markov Models. *RECOMB*, 68–75, 2004.

GAMOT: AN EFFICIENT GENETIC ALGORITHM FOR FINDING CHALLENGING MOTIFS IN DNA SEQUENCES

N. KARAOGLU, S. MAURER-STROH AND B. MANDERICK
Computational Modeling Lab and SWITCH Lab
Vrije Universiteit Brussel
Pleinlaan 2, 1050 Brussels
E-mail: {nkaraogl, smaurers, bmanderi}@vub.ac.be

Weak signals that mark transcription factor binding sites involved in gene regulation are considered to be challenging motifs. Identifying these motifs in unaligned DNA sequences is a computationally hard problem which requires efficient algorithms. Genetic Algorithms (GA), inspired from evolution in nature, are a class of stochastic search algorithms which have been applied successfully to many computationally hard problems, including regulatory site prediction. In this paper, we propose GAMOT, an efficient GA for solving Planted (l, d)-Motif Problems as introduced by Pevzner and Sze. We show empirically that our algorithm is not only able to solve the challenging problem instances with short motifs such as (14,4) and (15,4) efficiently but also that it is able to solve problems with longer motifs such as (20,7), (30,11) and (40,15). GAMOT can find the planted motifs in near-linear computational time thanks to an additional step which creates a highly fit population of solutions even before the evolutionary process is applied. We present a comparison of our results with some of the state-of-the-art algorithms such as VAS and PROJECTION.

1. Introduction

Weak signals in the genome regulating transcription of neighboring genes by representing a binding site for a transcription factor can be very challenging motifs to be identified. Efficient algorithms are needed for the computationally hard task of finding these motifs in unaligned DNA sequences. Pevzner and Sze [17] gave a combinatorial description of the problemwhich is known as the Planted (l, d)-Motif Problem (PMP) and proposed a challenge:

Planted (l, d)-Motif Problem: Let M be a fixed but unknown nucleotide sequence of length l (the motif consensus). Given t nucleotide sequences of length n each containing a variant of M with alterations at maximum d points, determine positions of the motif in each sequence.

Their challenge problem, shortly (15,4), involves finding a motif of length 15 with exactly 4 point mutations in 20 DNA sequences with length 600. This problem is hard since the signal is too weak for applying probabilistic methods while exhaustive search is impractical since the motif is too long [19][23]. Pevzner and Sze [17] proposed the WINNOWER algorithm which constructs a graph with vertices corresponding to substrings from the sample sequences and edges between similar substrings to tackle the challenge problem. Buhler and Tompa [4] analyzed the problem and showed that there is a certain threshold for maximum allowed mutations d for every parameterization of the problem with l, n and t such that no algorithm can distinguish the planted motif from patterns that occur randomly in the sequences. For example a random motif of length $l = 15$ matches a fixed motif with probability 1.2×10^{-4} when $d = 4$ substitutions allowed while it has a probability of bigger than 0.05 matching probability when $d = 8$ substitutions allowed. Using this observation Buhler and Tompa introduced additional challenging problem instances such as (9,2) (11,3), (14,4) and (18,6) where the likelihood of finding a true motif in the sequences is very small. They proposed an algorithm, PROJECTION, which is designed to efficiently solve challenging problem instances of the planted (l, d)-motif model using random projections. Other algorithms proposed for motif finding problems include exhaustive search algorithms [3][19][23][25] such as Sagot's suffix tree [18], Pavesi et. al's WEEDER algorithm [13] and Eskin et al's MITRA [6] and heuristic based algorithms such as Hertz and Stormo's CONSENSUS [11] and Bailey and Elkan's MEME [2]. Although many of the algorithms in the literature provide good results for short motif sizes, usually their running times or space requirements are exponential or they do not always guarantee the finding a solution.

Recently, Leung and Chin [7][8] proposed algorithms based on Buhler and Tompa's idea of random projections that can solve problems with even longer motifs such as (20,7), (30,11) and (40,15) which are considered intractable by many algorithms. Their basic voting algorithm [7] takes all motifs of length-l from the input set and calculates the subset of all variants with d point mutations and calculates the occurrences of these variants in the input set with the help of two very large hash tables. The length of the hash tables increase exponentially with the length of the motifs in order to avoid collisions. Their VAS [8] algorithm solves this by taking projections of length-l' from the length-l motifs in order to make finding longer motifs feasible. There are a few issues with their approach. Firstly, although it is improved, the time complexity of the algorithm is still exponential $O(nfl(nt)^k(4^{k-1}+1/4^{k-1})l)$ as well as the space complexity $O((nt)^k(4^{k-1}+1/4^{k-1})l)$. Secondly, the solution quality is reduced because the algorithm works with shorter projections of long motifs instead of the whole motif. Finally, the algorithm's performance depends highly on selection of the hash function and the right hash table size. Unfortunately, no hash function for motifs can guarantee a collision free hashing with shorter keys because the motifs are considered to be random in planted motif-(l, d) problems.

In this paper, we propose an efficient genetic algorithm (GAMOT) for solving challenging motif problems with long motif sizes such as (20,7), (30,11) and (40,15) with near linear time complexity and linear space complexity. We compare our results with exhaustive searches, previous applications of GA as well as some of the state-of-the-art algorithms such as VAS and PROJECTION and show empirically that our algorithm is able to solve challenging problem instances efficiently.

2. GA for Motif Finding

Genetic Algorithms (GA) are stochastic search algorithms inspired from evolution in nature, which have been applied successfully on many computationally hard problems including regulatory motif finding [1][20]. The application of GA in the task of regulatory signals identification has been first studied by Stravrovskaya and Mironov [20].

Approaches for attacking the motif finding problem can be grouped into two categories based on their representation of the motifs [5]. The first group of algorithms using position-specific scoring matrix (PSSM) representation tries to obtain a generative probabilistic representation of over-represented signals with frequencies of all nucleotides for each motif position retrieved from an alignment of the motifs. The second group of algorithms is based on a consensus pattern which comprises only the most frequent nucleotide on each motif position, but has the advantage of a simple string representation.

Stravrovskaya and Mironov [20] introduced GAs based on both these representation formulations of the problem widely used by many other algorithms.

In their first GA (GA1) based on PSSM, a chromosome represents an alignment of all sequences. The algorithm tries to find all alignments, which maximize the sum of the maximum frequencies of nucleotides at each position of the alignment over the motif length. A data set with t sequences with sequences of length n will contain $(n-l)^t$ alignments of motifs of length l.

In the second GA (GA2) the chromosome represents a candidate consensus string of length l and the algorithm tries to maximize the score of this consensus string with respect to all sequences in the data set. For a consensus string of length l, the algorithm will have to look for the best one out of 4^l possible motifs. It is clearly seen that the second algorithm will have to operate in a smaller search space for bigger values of t and n.

3. An Efficient Algorithm (GAMOT)

In this paper we propose an algorithm similar to GA2 above with an additional fast motif discovery step to obtain a highly fit set of initial solutions and special exploration operators to explore the search space of motifs efficiently.

3.1. *Fast motif discovery*

The drawback of exhaustive search algorithms is that they try all possible motifs of length l, which means that they look for the consensus string in 4^l motifs. However, if there actually is a motif to be found in a set of sequences, it should be inevitable to encounter motifs that are at least highly similar to the consensus already in the pool of sequences where the motif is expected. Consequently, only considering motifs occurring in the given sequences would result in a search space of $(n-l)t$ instead of trying all 4^l motifs. Even with mutations, planted motifs will have a better score than most of the motifs out of 4^l.

A possible algorithm to select motif candidates is given in Algorithm 1. The FAST MOTIF DISCOVERY algorithm takes a set DNA which contains t sequences of n base pairs with planted motifs of length l with d point mutations and returns a list of candidate motifs with fitness values above a certain threshold. The basic algorithm goes through all t sequences in the set, removing one at a time from the set, calculates fitness values of all $(n-l)t$ occurrences of strings of length l in the removed sequence with regard to the remaining strings in the set. However, the algorithm presented in Algorithm 1 selects N motifs randomly out of all sequences to reduce the time and space complexity of this operation as when N gets closer to the total length of the sequences, the number of planted motifs in the selected motifs will converge to their common expectation. The $Fitness$ function in the algorithm, in its simplest form is selected as $TotalDistance$ of the candidate motif to the remaining sequences in the set, which is defined as the smallest Hamming distance between string m among all possible choices of starting points s in all DNA sequences. In the best case, one of the variants of the actual motif will already have the smallest total distance and be returned in the list of candidate motifs by FAST MOTIF SEARCH algorithm.

Nevertheless, this approach will have difficulties in the case of challenging problem instances where there might be arbitrary strings of length l in a given sequence which have a smaller distance than the planted motif with mutations. Therefore, this algorithm alone does not guarantee an optimal solution. However, it can be useful to generate a set of good candidates for a local search algorithm such as GA.

Algorithm 1 Fast Motif Discovery

```
procedure FAST MOTIF DISCOVERY(DNA, t, n, l)
    List S
    for k ← 0 to t do
        sequence ← Sequence(DNA, k)                  ▷ pick a sequence from DNA
        dna ←DNA - sequence                          ▷ remove the sequence from the set
        for i ← 0 to N, t ≤ N ≤ n − l do
            candidate ← Motif(sequence, s, l)        ▷ get motif at position s
            fitness ← Fitness(candidate, dna)
            if GoodEnough(fitness) then
                INSERT candidate, S
            end if
        end for
    end for
    return S                                         ▷ Return the list of candidate motifs
end procedure
```

At the cost of $O(ntl)$, the Fast Motif Discovery algorithm provides an excellent solution fitness even before the evolutionary process. In contrast to other approaches, our genetic algorithm reconstructs the original motif string from this initial set of good candidates using the evolutionary operators rather than blindly trying out all positions, which maximize the score or trying out all motifs [20]. It is possible to take all of the strings occurring in the sequences as the initial set instead of the ones with good fitness values. While this might increase the quality of solutions in some cases, it will also increase the time for the genetic algorithm to weed-out the unsuitable ones inside the population. Thanks to the exploration operators GAMOT guarantees that the motif is found even if an instance of the planted motif does not exist in the initial set as a whole or in part.

3.2. *The genetic algorithm*

The algorithm is based on the steady-state GA proposed by Syswerda [21] using an additional step which creates an initial population with a very high fitness instead of generating random individuals.

GAMOT algorithm given in Algorithm 2 takes a set DNA containing t sequences of n basepairs and returns the consensus string of length l, which is used for determining the positions of the planted motifs in the sequences.

The algorithm starts with applying FAST MOTIF DISCOVERY to the given problem, which returns a list of strings with good fitness values occurring in all of the t sequences.The strings with the highest scores are then taken as the initial population, which guarantees to start with a highly fit population.The number of strings to be taken as the initial population depends on the population size parameter of GA which is usually tuned depending on the problem instance [26]. In the most extreme case population size can be selected as the number of all possible l-mers in the sequence which is $(n - l)t$. This is still a smaller set than 4^t and the likelihood of the solution being closer to the strings in this restricted subset will be higher than the rest $4^t-(n-l)t$ strings in the search space. Unfit strings in this initial set will be automatically weeded out by the selective pressure of the evolutionary algorithm that gives more chance to the solutions with higher fitness to be considered for crossover, further causing the population to be dominated gradually by the fittest solutions at each generation. Selecting a smaller portion of this initial set as the population may speed-up the weeding out process while in highly challenging problem instances good candidates may also be weeded out. Nevertheless, GAMOT will still be able to find the good solutions even if some of the good candidates were removed from the initial set. In our implementation, FAST MOTIF DISCOVERY returns the motifs with scores which are in the upper $\%50$ percentile and we have chosen population sizes, which give good results for planted-(l,d) motif problem experimentally.

After creating the initial population, the algorithm selects two individuals from the population using linear ranking [9][26]. Whitley [26] has shown that linear ranking gives better results compared to proportional selection. The algorithm then creates a new candidate consensus string using two-point crossover and replaces the worst individual with the newly created individual.

In order to explore the whole solution space, the algorithm applies designated exploration operators. The first operator is the MUTATE operator which is applied at random intervals by mutating some of the positions with a random nucleotide in an arbitrary individual selected from the population. We use a second operator, SHIFT, which shifts the nucleotides to left or right by one position to achieve a better alignment. The position opened by shifting is filled with a randomly selected nucleotide. This operator enables the algorithm to take advantage of candidate motifs if even only a portion of the motif exists in the string.

Algorithm 2 GAMOT

procedure GAMOT(DNA, t, n, l)
    $i \leftarrow 0$
    $S \leftarrow$FAST MOTIF DISCOVERY(DNA, t, n, l) ▷ Collect a good candidates
    INIT-POPULATION (S)
    repeat
        $C \leftarrow SELECT - PARENTS(P_i)$
        $Z \leftarrow RECOMBINE(C)$ ▷ try reconstructing the original motif
        MUTATE(Z) or ▷ Explore
        SHIFT(Z)
        $i \leftarrow i + 1$
        UPDATE-POPULATION (P_i, Z)
    until STOP() ▷ Fitness is satisfactory or i reached max. evaluations
    return BEST(P_i)
end procedure

The algorithm performs the steps discussed above until the score of the best individual is satisfactory or a certain number of evolutions are completed and returns this as the solution. In case it is not possible to make an initial estimation about the target fitness, it is possible to select a stop condition such as stopping when the fitness of the population does not improve for some generations.

After GAMOT has returned the consensus string with the best score, additional methods can be applied to find the transcription factor binding sites in the given sequences. The simplest method involves scanning all of the sequences and taking the start position of the best motif per sequence that has the minimum Hamming distance to the consensus string. For multiple motifs per sequence or no motif at all, a threshold based on the Hamming distance can be defined.

4. Experimental Results

We compare the performance of GAMOT with exhaustive search methods, with the two other GA described earlier, as well as some state-of-the-art algorithms based on other methodologies using randomly generated test data.

We have generated 20 random DNA sequences of length 600 and planted a random motif of length l with exactly d random mutations to the sequences for different values of l and d. The data sets are generated using different seeds than the GA with the state-of-the-art Mersenne Twister random number generator [16] which guarantees equidistribution in 623 dimensions to avoid bias in the experiments. The datasets used in the experiments

Table 1. Comparison with exhaustive search and other GA.

l	d	$GAMOT$	$GA2$	$GA1$	$BBBM$	MSS
8	2	13s	34s	47s	19s	99s
10	2	15s	41s	17min	7min	33min
12	3	12s	318s	36min	165min	11hrs
14	4	10s	18.2min	-		
16	5	13s	37.1min	-		

are publicly available and can be obtained via email from the corresponding author. The measurements are made on an IBM xSeries 330 computer with Intel Pentium III processor and 1GB of RAM running Windows XP operating system.

4.1. *Comparison with exhaustive search*

We have implemented two exhaustive search algorithms as described in Jones and Pevzner [14]. The first algorithm is the brute-force Median String Search (MSS) which tries all possible motifs of length l and finds the one with the minimal total distance. The second algorithm we use for benchmarking is an improved version of the Branch and Bound Median String Search Algorithm (BBBM). This algorithm speeds up the search Median String Search by cutting the branches in the search tree which contain motifs that have total distances worse than the best solution found so far. In our version we have also added a step to make a reasonable initial estimation by random guess instead of starting with infinite distance to the solution.

We have measured the running time of each algorithm to find the exact solution on several problem instances. Algorithms stop when the score of the best individual is equal to the score of the planted motif. Table 1 shows the average runtimes of the algorithms. GAMOT is able to solve the given problems even for (14,4) and (16,5) problems which are infeasible to compute with exhaustive search algorithms like MSS and BBBM.

4.2. *Comparison with GA1 and GA2*

We have also implemented GA1 and GA2 as discussed before (see section 2) andcompared solution performance as given in Table 1 GAMOT finds the planted motif faster than GA1 and GA2 while these algorithms are faster than exhaustive search methods used in our comparison. GA2 gives better results than GA1 as it operates in a smaller search space.

4.3. *Comparison with other algorithms*

In the second set of experiments we have compared GAMOT with state-of-the-art algorithms in the literature using the same problem instances. Table 2 and Table 3 show the average runtimes of the algorithms.

GAMOT is able to find the solution faster than PROJECTION, however VAS is much faster in returning solutions in problem instances with short planted motifs. However, for

Table 2. Evaluation speeds for short motifs.

l	d	$GAMOT$	$PROJECTION$	VAS
8	2	13s	31s	1s
10	2	15s	36s	1s
12	3	12s	138s	3s

Table 3. Evaluation speeds for longer motifs.

l	d	$GAMOT$	$PROJECTION$	VAS
14	4	10s	6min	13min
16	5	13s	13min	25s
18	6	14s	18min	2min
20	7	27s	26min	6min
30	11	59s		98min
40	15	10min		12min

the problem instances with longer motifs GAMOT finds the solution faster than the others. Moreover solution time of GAMOT increases almost linearly as the planted motif size gets longer.

However, Table 2 and Table 3 shows only one side of the story since some of the algorithms in the literature can only find good approximations to the actual solution. We have also compared the quality of the solutions using the performance coefficient defined by Pevzner and Sze [17].

4.3.1. *Quality of the solutions*

Let K denote the set of t l-base positions in the t occurrences of a planted motif, and let P denote the corresponding set of base positions in the t occurrences predicted by an algorithm. Then the algorithm's performance coefficient on the motif is defined $K \bigcap P / K \bigcup P$. When all occurrences of the motif are found correctly, the performance coefficient achieves its maximum value of one.

The quality of the solutions is shown in Table 4. GAMOT finds always the exact solution; therefore, it has a performance coefficient of 1 in all problem instances. It is clearly seen that GAMOT is able to find the solution in every problem instance within the given times whereas other algorithms produce solutions with poor quality as the problem complexity increases.

4.4. *GAMOT parameters*

GA are generic search algorithms, which can be applied to any problem domain. Efficiency of a GA higly depends on tuning its parameters for a problem domain or instance. It is known in the literature that the complexity of the problem and the length of the rep-

Table 4. Quality of solution $K \bigcap P / K \bigcup P$.

l	d	$GAMOT$	$PROJECTION$	VAS
10	2	1	0.67	0.86
11	2	1	0.9	0.18
12	3	1	0.9	0.83
13	3	1	0.82	0.87
14	4	1	0.9	na
15	4	1	1	1
16	5	1	0.74	0.65
17	5	1	0.74	0.94
18	6	1	1	0.86
20	7	1		0.73
30	11	1		1
40	15	1		1

resentation have a critical role in determining the population size [10], which seems to be the most influential parameter on the performance of a GA. We have run a series of experiments for determining suitable problem sizes for the GAMOT algorithm. Our results are shown in Figure 1.

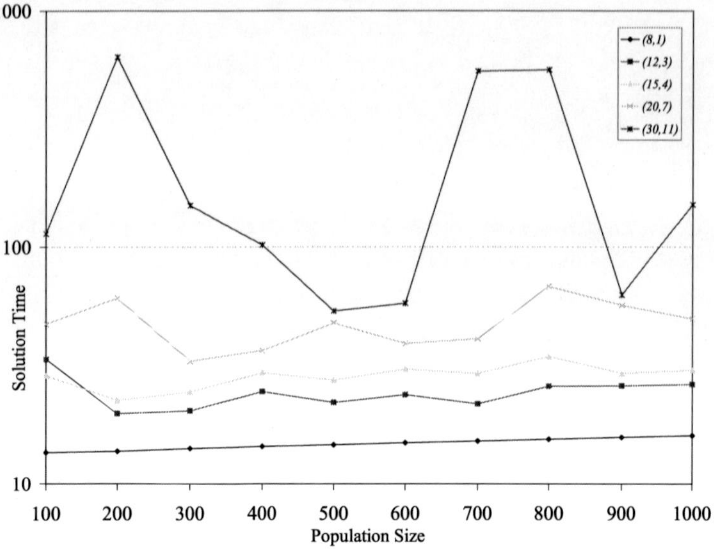

Figure 1. Population sizes for different planted-(l,d) motif problems and algorithm performance.

We see that when problem complexity is small as in (8,1) problem, the population size can be selected as small as 100 individuals. As the problem complexity increases bigger

Table 5. Table 4.5 GAMOT parameters.

$Parameter$	$Value$
$PopulationSize$	$100 - 700 for (8, 1) - (40, 15)$
$SelectionPoolSize$	0.80
$IneritanceLevel$	0.90
$ExplorationProbability$	0.35
N	$0.5t(n - l)$

population sizes become more suitable as in (30,11) problem where a population size of 500 gives good results. In our experiments we have used the optimal values as shown in the above graph. A summary of the GAMOT parameter values used in our experiments is given in Table 5.

In order to increase selective pressure, selection operator considers the motifs with fitnesses (parameter $SelectionPoolSize$), which are in the upper $\%80$ percentile enabling GAMOT to converge to the solution faster.

The parameter $InheritanceLevel$ of GAMOT, which controls the number of individuals to be kept from previous generation is set to $\%90$ when obtaining the results presented in this paper.

Furthermore, the exploration operators of GAMOT are applied with a probability of 0.35 (parameter $ExplorationProbability$) in our experiments. The parameter N for the fast motif discovery step is set to $0.5t(n - l)$,where t is the number of sequences, n is the length of a sequence and l is the motif size. The parameters presented here are given for informational purposes. Better values can be found for different problem instances.

5. Conclusions and Future Work

In this paper, we have proposed an efficient Genetic Algorithm for the Planted Motif Problems and shown empirically that it is always able to find the exact solution even in very complex problem instances (18, 6), (20, 7), (30, 11) and (40, 15) where the other algorithms perform poorly. Secondly, we have shown that overall solution times of the algorithm are much smaller than other algorithms given in the literature without having to compromise from the space complexity in favor of time complexity. Apparently, GAMOT can find the planted motifs in near-linear computational time, which would qualify the method also for large-scale motif identification projects, even for longer motifs. The proposed algorithm achieves these results firstly because it uses a smaller search space based on consensus strings. Secondly, because of an additional step which creates a highly fit population of solutions by identifying candidate motifs inside the given sequences even before the evolutionary process is applied. It is interesting to note that the first restriction to the simpler consensus string method does not impact on the quality of the solutions, at least in the tested instances with simulated data.

GAMOT algorithm has two advantages over the existing motif finding algorithms. First and foremost, GAMOT algorithm produces better solutions. In our experiments GAMOT was able to achieve a performance coefficient of 1 in all of the test cases while the others failed to achieve the same in some of the test cases. Secondly, it produces good quality results in much shorter time than the other algorithms when it comes to challenging problem instances with long motif sizes. Furthermore, GAMOT does not suffer the exponential space and time complexity which VAS algorithm suffers. Clearly, the motif partitioning scheme to reduce time and space complexity in VAS algorithm causes the algorithm to sacrifice from the solution quality as well. GA are good candidates for computationally-hard problems and there are many flavors introduced in the literature which might produce a faster convergence to a solution. They are also quite suitable for parallel computation.

In this paper, only one flavor of GA is discussed, nevertheless it might be possible to achieve even better results with other GA. Furthermore, in many cases it is also advisable to tune the parameters of the GA to achieve better results. Our results have shown that GA with an additional step for selecting a good set of candidate solutions improves the performance of the simple GA. In this paper, we have used FAST MOTIF DISCOVERY with $TotalDistance$ as scoring function for this purpose. However, it is possible to replace or merge FAST MOTIF DISCOVERY with other methods such as Gibbs sampling or PROJECTION and apply the GA for refinement of the solution.

The GAMOT algorithm finds a good consensus string quickly and we have used this consensus string to identify the sites which have the smallest distance to the consensus string for finding the transcription factor binding sites. This scheme can be further extended by translating the alignment of sites identified by GAMOT into a PSSM, which can then be used for further identification of motifs in the initial query set of sequences (refinement of the predicted motif) or application to find similar sites in other upstream sequences using the presumably more sensitive PSSM approach.

The algorithm we present here produces good results on simulated data and can be further equipped with additional mechanisms suitable for real-life data to be used in finding transcription factor binding sites such as additional filters for known unspecific repetitive elements and a better scoring function. Leung and Chin [8] report good results using background sets. We would like to incorporate some of these methods for testing our algorithm on benchmark data sets, which recently became available [24].

References

1. Aerts, S., Van Loo, P., Moreau, Y. and De Moor, B., A genetic algorithm for the detection of new cisregulatory modeules in sets of coregulated genes, *Bioinformatics*, 20(12) 1974–6, 2004.
2. T. Bailey and C. Elkan, Unsupervised learning of multiple motifs in biopolymers using expectation maximization, *Machine Learning*, 21:51–80, 1995.
3. A. Brazma, I. Jonassen, I. Eidhammer and D. Gilbert, Approaches to the automatic discovery of patterns in biosequences, *Journal of Computational Biology* (JCB), 5:279–305, 1998.
4. J. Buhler and M. Tompa, Finding Motifs Using Random Projections, *Journal of Computational Biology* (JCB), 9(2):225–242, 2002.
5. Eleazar Eskin, From Profiles to Patterns and Back Again: A Branch and Bound Algorithm for Finding Near Optimal Motif Profiles, in Proceedings of the Eight Annual International Conference on Research in Computational Molecular Biology (RECOMB-2004).
6. Eleazar Eskin, Uri Keich, Mikhail S. Gelfand and Pavel A. Pevzner, "Genome-Wide Analysis

of Bacterial Promoter Regions," in Proceedings of the Pacific Symposium on Biocomputing (PSB-2003), Kaua'i, Hawaii: January 3–7, 2003.

7. Francis Y. L. Chin and Henry C. M. Leung, Voting Algorithms for Discovering Long Motifs, Proceedings of the Third Asia-Pacific Bioinformatics Conference (APBC2005), 261–271 (January 2005) http://www.cs.hku.hk/chin/paper/apbc05.pdf.
8. Francis Y. L. Chin and Henry C. M. Leung, "An Efficient Algorithm for String Motif Discovery", Proceedings of the Fourth Asia-Pacific Bioinformatics Conference (APBC2006), Taipei, Taiwan, (February 2006) (accepted).
9. D. E. Goldberg, Genetic Algorithms in Search, Optimization, and Maching Learning, Addison-Wesley, 1989, ISBN 0-201-15767-5.
10. D. E. Goldberg, K. Deb, and J. H. Clark, Accounting for noise in the sizing of populations, in L. D. Whitley, editor, Foundations of Genetic Algorithms 2, pages 127–140, Morgan Kaufmann, 1992.
11. Hertz, G. Z. and Stormo, G. D., Identifying DNA and protein patterns with statistically significant alignments of multiple sequences, *Bioinformatics*, 15(7–8):563–77, 1999.
12. J. H. Holland, Adaption in natural and artificial systems, University of Michigan Press, Ann Arbor 1975.
13. Pavesi G., Mauri G. and Pesole G., An algorithm for finding signals of unknown length in DNA sequences, *Bioinformatics*, 2001; 17 Suppl. 1:S207-14.
14. Jones, N. C. and Pevzner, P. A., An Introduction to Bioinformatics Algorithms, MIT Press, Cambridge, Mass., 2004.
15. U. Keich and P. A. Pevzner, Finding motifs in the twilight zone, in Proc. 6th Annual International Conference on Research in Computational Molecular Biology (RECOMB 2002).
16. M. Matsumoto and T. Nishimura, "Mersenne Twister: A 623-dimensionally equidistributed uniform pseudorandom number generator", ACM Trans. on Modeling and Computer Simulation Vol. 8, No. 1, January, pp. 3–30 (1998).
17. Pevzner, P. A. and Sze, S. H., Combinatorial approaches to finding subtle signals in dna sequences, The Eighth International Conference on Intelligent Systems for Molecular Biology, pp. 269–278, 2000.
18. M.-F. Sagot, Spelling approximate repeated or common motifs using a suffix tree, in C. L. Lucchesi and A. V. Moura, editors, LATIN'98: Theoretical Informatics, Lecture Notes in Computer Science, pages 111–127, Springer-Verlag, 1998.
19. Staden, R. (1989), Methods for discovering novel motifs in nucleic acid sequences, *Comput. Appl. Biosci.*, Vol. 5(5), pp. 293–298.
20. Elena, D. Stavrovskaya and Andrey A. Mironov, Two genetic algorithms for identification of regulatory signals, in Silico Biology 3: 5 (2003).
21. G. Syswerda, Uniform crossover in genetic algorithms, in J. D. Schaffer, editor, Proceedings 3rd International Conference on Genetic Algorithms, pp. 2–9, Lawrence Erlhaurn Associates 1989.
22. W. Thompson, E. C. Rouchka and C. E. Lawrence, Gibbs Recursive Sampler: finding transcription factor binding sites, Nucleic Acids Research, 2003, Vol. 31, No. 13 3580-3585.
23. Tompa M., An Exact Method for Finding Short Motifs in Sequences, with Application to the Ribosome Binding Site Problem, Proc. Int. Conf. Intell. Syst. Mol. Biol. 262–271, 1999.
24. Martin Tompa, Nan Lil, Timothy L. Bailey, George M. Church, Bart De Moor, Eleazar Eskin, Alexander V. Favorov, Martin C. Frith, Yutao Fu, W. James Kent, Vsevolod J. Makeev, Andrei A. Mironov, William Stafford Noble, Giulio Pavesi, Graziano Pesole, Mireille Regnier, Nicolas Simonis, Saurabh Sinha, Gert Thijs, Jacques van Helden, Mathias Vandenbogaert, Zhiping Weng, Christopher Workman, Chun Ye, Zhou Zhu, "An Assessment of Computational Tools for the Discovery of Transcription Factor Binding Sites," Nature Biotechnology, 23(1):137–44, 2005.
25. van Helden, J., B. Andre and J. Collado-Vides, Extracting regulatory sites from the upstream

region of yeast genes by computational analysis of oligonucleotide frequences, *J. Mol. Biol.*, 266:231–245, 1998.

26. Whitley D., ed. J. Schaffer, Proc. 3rd Int. Conf. on Genetic Algorithms (Fairfax, VA, June 1989), The GENITOR Algorithm and Selective Pressure: Why Rank-Based Allocation of Reproductive Trials is Best, San Mateo, CA: Morgan Kaufmann, 1989.

IDENTIFICATION OF SPACED REGULATORY SITES VIA SUBMOTIF MODELING

E.WIJAYA AND R. KANAGASABAI

Institute for Infocomm Research,
21 Heng Mui Keng Terrace,
Singapore 119613
E-mail: kanagasa@i2r.a-star.edu.sg

In this paper we propose a novel approach for identification of generic motifs in an integrated manner by introducing the notion of submotifs. We formulate the motif finding problem as a constrained submotif pattern mining and present an algorithm called SPACE for identifying motifs that may contain spacers. When spacers are present, we show that the algorithm can identify motifs where 1) the spacers may be of varying lengths, 2) the number of motif segments may be unknown, and 3) the lengths of motif segments may be unknown. We perform rigorous experiments with the Motif Assessment Benchmarks by Tompa et al., and observe that our algorithm overall is able to outperform all popular algorithms tested so far, with significant improvements on sensitivity and specificity.

1. Introduction

The availability of vast genomic sequences from several organisms provides us with a rich opportunity for advancing our knowledge about biological systems. However, regulatory networks in most of these sequenced genomes is largely unexplored. One of the major challenges facing biologists today is to understand the regulatory mechanism of genes. This challenge includes detection of transcription factor binding sites involved in regulation and discovery of regulatory networks.

The problem of *de novo* identification of transcription factor binding site motifs has been widely studied and a number of motif-finding algorithms have been proposed. Most of these algorithms can be classified as follows:

- profile-based methods that addresses motif finding by learning a matrix (profile) model that describes the binding sites, e.g. Gibbs sampler [13], MotifSampler [20], SeSiMCMC [6], MEME [2], NestedMICA [11], Improbizer [1].
- consensus-based methods that enumerate oligos of (or upto) a given length to find strings that appear in many sequences and use statistical measures to rank them, e.g. MULTIPROFILER [12], Weeder [17], MITRA [5], TEREISIAS [18], Gemoda [10], CMF [24].
- hybrid methods that use a combination of the above two methods e.g. [8].

In spite of the availability of these methods, motif-finding (in general) continues to be a difficult problem because of challenges like complex motif structure, presence of weak signals and scalability to the whole genome. For example, recently Tompa [21] conducted an assessment of 13 popular motif discovery algorithms over 56 datasets drawn from *H. Sapiens*, *M. Musculus*, *D. melanogaster*, and *S. cerevisiae* genomes, and found that all the algorithms performed badly overall. In fact, barring yeast datasets, the performance on all organisms was significantly worse. This motivates developing more sophisticated motif-finding algorithms.

In the recent survey Eisen [4] brought up the issue that many of these contemporary methods does not incorporate the structural properties of the binding sites. In the real biological context, the regulatory motifs may be more intricate. For example, they may appear as motifs with spacers with sparse conservation in the binding sites. This characteristics is significant in the regulatory mechanism, because two or more sites are often recognized by the same protein (as is frequently the case, for instances, of RNA polymerase) [14]. Moreover the binding sites are often recognized by different macromolecular complexes that make contact with one another [16,25]. Our focus in this paper is to find such spaced motifs.

2. Related Work

There have been some works on extraction of motifs with spacers. OligoDyad [22] finds spaced motifs by counting the number of occurrences of each defined spaced pair of trinucleotides, before finally assessing its statistical significance. MITRA [5] is another algorithm for finding dyads. SesiMCMC [6] first optimizes the weight matrix with Gibbs sampling for a given motif and spacer length, and finally looks for the best motif and spacers length based on the information content of motif and distributions of motif occurrence position. The methods reported in [3,15] use suffix tree to store the regularly spaced motif before finally identifying the motif pairs. YMF [19] handles spacers by enumerating all the possible gap lengths between two composite motifs.

The approaches used by the existing methods can be characterized in three ways. The first, and most common, approach is to address the problem by assuming fixed length spacers. The second approach is to address the finding of spaced motifs as finding single motif. The main characteristic of this approach is the use of efficient data structures to traverse the space of IUPAC patterns. The third approach handles spacers by enumerating all the possible gap lengths between two composite motifs. Though this approach can find motifs with varying length spacers, it is computationally expensive and often limited to finding short motifs.

In this paper we propose a new approach for finding spaced motifs, and develop a novel motif-finding algorithm that offers flexibility in handling variations in spacer length, the number of motif parts (henceforth called *motif segments*) and their lengths.

3. Our Approach

Central to our approach is the notion called the *submotifs*. Submotif is defined as a conserved sub-region of a given motif (spaced or not). Our idea is to use a divide-and-conquer approach whereby we find the target motif by first finding its submotifs and then strategically compositing them to deduce the target motif. The number of submotifs is assumed to be unknown and they may be possibly overlapping. This setting has the following advantages:

(1) The length of the target motif need not be known *even if* we pre-fix the length of the submotif. This follows because any l-mer can be represented as a union of its substrings (possibly overlapping) of length $l_s, l_s < l$. Thus this method can result in higher sensitivity.
(2) The conservation of the instances formed by union of the submotif instances will be stronger in conservation than the instances obtained directly without submotifs. This can be seen by observing that the former instances will always be a subset of the latter. This implies that the method can yield higher specificity.
(3) It provides a natural extension for finding motifs with spacers, in which neither the spacer length nor the segment length need to be known. This is because the method attempts to find the longest motif and the presence of spacers does not preclude submotif finding (provided the compositing of submotifs is done appropriately).

However, the challenge is that there could be potentially too many submotifs (many of them spurious) and how effectively the submotif-compositing can be done to return "good" motifs. To tackle this problem, we formulate this task as a constrained frequent itemset mining problem and propose a new algorithm for solving it. This algorithm can output spaced patterns optimized based on user considerations such as degree of conservation, range of spacer widths, etc. The algorithm and our overall method are described in the following sections.

The rest of the paper is organized as follows. Section 4 explains the definition of our novel motif model and terminologies. It is followed by section 5 where we describe our proposed method in detail. In section 6 we present our experimental results on benchmark dataset in comparison with existing tools and results on real biological data. Finally in section 7 we will conclude our paper by discussing the strengths and limitations of our approach and directions for our future research.

4. Problem Definition

Let $\mathbf{S} = \{s_1, s_2, \ldots s_t\}$ be a set of t sequences. Our aim is to identify the target motif(s) and its instances from this set of sequences. We view the target motif as a cascade of segments separated by spacers, where both the segments and spacers could be of varying lengths (that is unknown). Thus we define the motif finding problem as one of finding all segments and spacers of a motif. As stated in the previous section, our key concept is *submotif* which will be used to model the segments and in turn the motif.

Let $hd(x, x^{'})$ denote the Hamming distance between strings x and $x^{'}$. A string x is called a $(\mathbf{l_s}, \mathbf{d})$- **submotif** of a set of strings **X'** $= \{x^{'}_1, x^{'}_2, \ldots, x^{'}_k\}$, each of length l_s if $hd(x, x^{'}_i) \leq d; \forall i = 1, 2, \ldots, k$. In this case $x^{'}_i$ is called an instance of the $(\mathbf{l_s}, \mathbf{d}) -$ **submotif** x. Next we define the motif segments.

A string m is called a $(\mathbf{l_s}, \mathbf{d}) -$ **segment** of length $|m|$ if it contain at least 2 overlapping $(\mathbf{l_s}, \mathbf{d})$- **submotifs**. The total length of the $(\mathbf{l_s}, \mathbf{d})$-submotif is equal to $|m|$. For illustration, consider a 3-segments spaced motif shown in Figure 1. In the example, x_{11} and x_{12} are (l_s, d)-submotifs with x'_{11} and x'_{12} as their instances. Furthermore m_1 is a (l_s, d)-segment with m'_1 as one of its instances.

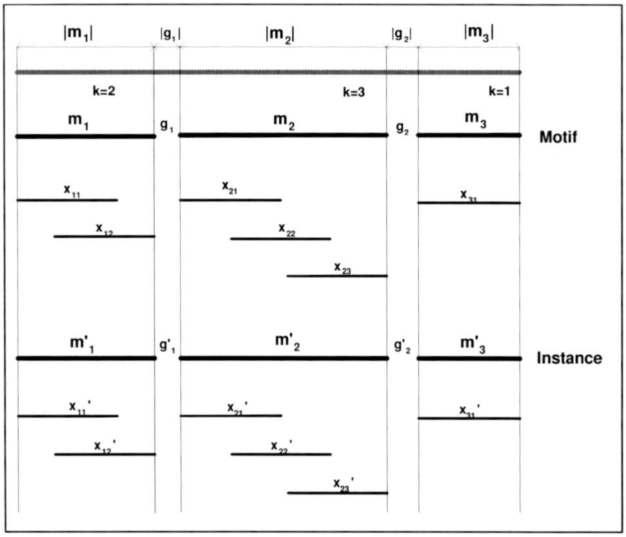

Figure 1. A 3-parts motif with submotifs and an instance.

A string $M = m_1 \cdot g_1 \cdot m_2 \ldots m_{N-1} \cdot g_{N-1} \cdot m_N$ is called a $\mathbf{(l_s, d, q, e) - motif}$ of **S**, where m_i is a string for $i = 1, 2 \ldots, N$ and g_i is a spacer of length $|g_i|$ for $i = 1, 2, \ldots, N-1$. And it has the following properties:

(1) there exist at least q of its instances $M' = m'_1 \cdot g'_1 \cdot m'_2 \ldots m'_{N-1} \cdot g'_{N-1} \cdot m'_N$ in **S**.
(2) m'_i is a $(l_s$,d)-segment instance of m_i for $i = 1, 2 \ldots, N$.
(3) $|g_i - g'_i| \leq e$ for $i = 1, 2 \ldots, N$.

For illustration, in Figure 1, $M = m_1 \cdot g_1 \cdot m_2 \cdot g_2 \cdot m_3$ is a $\mathbf{(l_s, d, q, e) - motif}$. Given a set of sequence **S**, our motif finding problem is to find $\mathbf{(l_s, d, q, e) - motif}$. In the next section, we describe our algorithm, called SPACE to solve this problem.

5. Algorithm SPACE

Algorithm SPACE consists of three major components namely, the *generation of candidate patterns*, *constrained frequent pattern mining*, and finally *significance testing*. Generation of candidate patterns is the step where we list the candidate motifs and their respective (l_s, d)-segments and (l_s, d)-submotifs. In the second step, we process every candidate motif and its segments and submotifs to mine frequent submotif patterns. Eventually, in the last part we score the motifs based on its input sequences and background model significance testing. Figure 3 depicts the overall strategy of our algorithm.

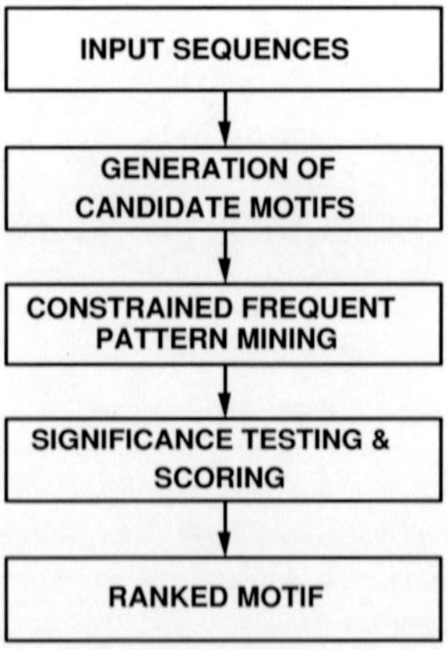

Figure 2. Outline of SPACE motif discovery algorithm.

5.1. *Generation of candidate motifs*

In this step, a set of motif candidates $\mathbf{M} = \{M_1, M_2, \ldots, M_k\}$ for all $k = 1, 2, \ldots, N \leq t(n - l + 1)$ is identified from the input sequences $\mathbf{S} = \{s_1, s_2, \ldots s_t\}$. This is done by enumerating all l length substrings from $\mathbf{S}$ and then scanning the sequences to obtain instances for each candidate motif generated. Motif candidates M_i will be formed by (l_s, d)-submotif. Similarly its instances will be formed by string of length l_s where $l_s < l$. And Hamming Distance between (l_s, d)-submotif with its instances is less or equal to d. Furthermore we require that the coverage of the submotif instances should be greater or equal than l_c. The coverage value l_c determines in what way does the conserved segments are placed to each other. This measurement allows us to determine segments that comes with spacers of varying length. In principle the l_c can be seen as the total length of the N number of (l_s, d)-segment strings of the given instances. It is computed as follows: $\sum_{i=1}^{N} |m'_i|$.

At the end of the candidate motif generation step, we will have a hash table $\mathbf{B}$ in which each entry is a motif candidate and its corresponding instances. Note that each instance in turn consists of submotif instances, some of which could be spurious. To minimize the spurious instances and hence maximize the chances of identifying the target motif, we employ a frequent motif pattern mining method.

5.2. *Constrained frequent pattern mining*

We present a novel method that aims to identify genuine submotifs (and hence the segments) borrowing ideas from Association Rule Mining [9]. However, it should be noted that our intention is not to mine rules but rather submotif patterns that are more *structured.* The mined patterns are structured because 1) the order of the submotifs is important, 2) the spacings between them have to be uniform. These two conditions imply a constrained association mining algorithm, described below.

Before presenting the algorithm, we introduce a key concept called the *generalized gap.*

This is used to denote the spacings between the submotifs which could overlap[a].

5.2.1. *Generalized gap*

Gap between two strings is conventionally described as any maximal, consecutive run of spaces in a single string [7]. This description presupposes that they are non-overlapping. We hereby introduce a new notion of a generalized definition of gap that includes the overlapping situation.

Generalized gap 5.1. Let p be the position of x_1 and q be the position of x_2 in a string M. We say that (p, q, α) is a pair of a string M if $\alpha = M[p..p+|\alpha|-1] = M[q..q+|\alpha|-1]$, and $p < q$. The gap of pair x_1 and x_2, denoted as $g(x_1, x_2)$ is the number of characters $q-p-|\alpha|$ between the two occurrences of the substring α. If the gap is non-negative then the pair is non-overlapping.

The following lemma describes the additive property of the generalized gap, which will be useful in our analysis later in this section.

Lemma 5.1. *Given an ordered substrings set* $\mathbf{x} = \{x_1, x_2, \ldots, x_K\}$ *of string* w*, each of length* l_s*. The generalized gap between* x_1 *and* x_K *is:* $g(x_1, x_K) = g(x_1, x_2) + \ldots + g(x_{K-1}, x_K) + (K-2) * l_s)$.

For the illustration of generalized gap and proof of the lemma above please refer the full paper version.

5.2.2. *Mining of constrained frequent patterns*

Recalling from section 5.1, the generation of candidate motifs processes the input sequences to output a table **B**. Note that each entry of this table is a motif candidate and its instances, and each motif candidate instance in turn contains (l_s, d)-submotif instances. The constrained frequent pattern mining step processes each entry of the table **B** and generates frequent motif patterns. We describe this process by considering a single entry of this table, i.e. a motif candidate M and its instances. This set of instances can be viewed as a *transactional* database, henceforth called **D**.

Let $\mathbf{M'} = \{M'_1, M'_2, \ldots, M'_i\}$ be a set of instances of M, for some i. Let $\mathbf{X'_i} = \{x'_{i1}, x'_{i2}, \ldots, x'_{ij}\}$ be the set of (l_s, d)-submotif instances contained in M'_i, for $j = 1, \ldots l - l_s + 1$. Then, **D** can be thought of as a table with M'_i as its rows, as shown in Table 1 below.

We first map all the submotif instances $\mathbf{X'_i}$ in **D** to a distinct set of items, called I. Given $\mathbf{M'} = \{M'_1, M'_2, \ldots, M'_i\}$ and $\mathbf{X'_i} = \{x'_{i1}, x'_{i2}, \ldots, x'_{ij}\}$, let the union of $\mathbf{X'_i}$, namely $\mathbf{X'} = \{\mathbf{X'_1} \cup \mathbf{X'_2} \ldots \cup \mathbf{X'_i}\}$. Then we cluster the elements of $\mathbf{X'}$ if the pairwise Hamming distance between any two strings is less or equal to $2d$. Elements of $\mathbf{X'}$ that do not satisfy this criteria are assumed to form isolated clusters. Let there be k clusters formed using this procedure. We reduce each cluster to a string that is a consensus of the strings in the cluster.

[a]We use 'gap' and 'spacer' interchangeably to denote the same concept.

M'_i	$\mathbf{X'_i}, \mathbf{G'_i}$
M'_1	$x'_{11}, g(x'_{11}, x'_{13}),\ x'_{13}$
M'_2	$x'_{23}, g(x'_{23}, x'_{25}),\ x'_{25}, g(x'_{25}, x'_{26}),\ x'_{26}$
M'_3	$x'_{32}, g(x'_{32}, x'_{33}),\ x'_{33}$
...	...
M'_i	$x'_{ia}, g(x'_{ia}, ...), \ldots, g(..., x'_{ib}), x'_{ib}$ s.t. $(1 \leq (a, b) \leq i)$

Denoting the consensus by x_i, we obtain the set $\mathbf{I} = \{x_1, x_2, \ldots, x_k\}$. We will call this as the submotif set.

Using the submotif set $\mathbf{I}$, we will derive frequent submotif pattern(s) in $\mathbf{D}$. We will deduce the motif(s) from these frequent pattern(s). We provide some definitions before describing our method.

Definition 5.2. (n-itemset pattern) A tuple $\mathbf{P_n} = (x_1, g_1, x_2, g_2, \ldots, g_{n-1}, x_n), n \geq 2$, is called a n-itemset pattern, where $x_i \in I$ and $g_i = g(x_i, x_{i+1})$, is the generalized gap between x_i and x_{i+1} as defined in Definition 1. For $n = 1$, $\mathbf{P_1} = (x_1)$.

Unlike the definition of itemset in Association Rule mining, the itemset in our algorithm also includes the integer value of a gap.

Since x_i is a substring of a motif candidate (which is of length l), observe that the biggest gap occurs when there are only two strings x_1 and x_2 and they are located at the two ends of the motif candidate. Then the gap between them is $l - 2l_s$. We denote this maximum gap as g_{max}. The smallest gap occurs when two adjacent substrings x_i and x_{i+1} overlap such that the gap is $-l_s + 1$. (Note that they cannot overlap entirely, because then x_i and x_{i+1} will be interpreted as a single submotif instance. See step 8 in Algorithm 1.) We denote this minimum gap as g_{min}.

Definition 5.3. (Length of n-itemset pattern) Given a n-itemset pattern $\mathbf{P_n}$, the length of $\mathbf{P_n}$ is defined as $L(\mathbf{P_n}) = g(x_1, x_n) + 2l_s$.

We next describe a tree growing procedure that will be useful in the description of our algorithm.

Consider a tree $\mathbf{T}$. We assume the tree has an empty root node at Level 0. We will grow the tree such that the nodes in the n-th level of the tree are n-itemset patterns, for $n \geq 1$. Level 1 nodes also can be thought of as 1-itemset patterns one each for every x_i. i.e. if the k-th node in Level 1 is denoted $\mathbf{P_1^{k_1}}$, then $\mathbf{P_1^{k_1}} = (x_{k_1}), k_1 = 1, ..., k$. Each Level 1 node is grown as follows. Let $\mathbf{G} = \{g_{min}, \ldots, g_{max}\}$. Let $\mathbf{L_2^{k_1}}$ stand for the children of $\mathbf{P_1^{k_1}}$. Then $\mathbf{L_2^{k_1}} = \mathbf{P_1^{k_1}} \times \mathbf{G} \times \mathbf{I}$, where: $\mathbf{P_1^{k_1}} \times \mathbf{G} \times \mathbf{I} = \{(\mathbf{P}, g, x) | \mathbf{P} \in \mathbf{P_1^{k_1}}, g \in \mathbf{G},\ and\, x \in \mathbf{I}\}$.

It can be noted that each element of $\mathbf{L_2^{k_1}}$ is actually a 2-itemset pattern denoted by $\mathbf{P_2^{k_1,k_2}} = (x_{k_1}, g_1, x_{k_2})$, where $g_1 \in G$, $x_{k_1} \in I$. Generalizing this process to the n-th level, we obtain: $\mathbf{L_n^{k_1,\ldots,k_{n-1}}} = \mathbf{P_{n-1}^{k_1,\ldots,k_{n-2}}} \times \mathbf{G} \times \mathbf{I}$.

Each element of $\mathbf{L_n^{k_1,\ldots,k_{n-1}}}$ is a n-itemset pattern, given by $\mathbf{P_n^{k_1,\ldots,k_n}} =$

$(x_{k_1}, g_1, \ldots, g_{n-1}, x_{k_n})$, $x_{k_i} \in \mathbf{I}$ and $g_i \in \mathbf{G}$. For convenience we denote $\mathbf{P_n^{k_1,\ldots,k_n}}$ by $\mathbf{P_n^{k_n}}$. The tree growth is stopped at node $\mathbf{P_n^{k_n}}$ if $L(\mathbf{P_n^{k_n}}) = l$.

Before describing our algorithm, we provide some definitions:

Definition 5.4. (Instance Satisfaction) Given a n-itemset pattern $\mathbf{P'_n} = \{x'_1, g'_1, x'_2, g'_2, \ldots, g'_{n-1}, x'_n\}$, where $\mathbf{P'_n}$ is a subset of $\mathbf{SM'_i}$. We define $\mathbf{X'_n} = \{x'_1, x'_2, \ldots, x'_n\}$ and $\mathbf{G'_n} = \{g'_1, g'_2, \ldots, g'_{n-1}\}$ such that $\mathbf{X'_n}$ and $\mathbf{G'_n}$ is a subset of $\mathbf{P'_n}$. And given a gap tolerance value e. We say that $\mathbf{P'_n}$ is the valid occurrence of $\mathbf{P_n}$ if all the following conditions are satisfied:

1) $|\mathbf{P_n}| = |\mathbf{P'_n}|$,
2) $hd(x_i, x'_i) \leq d$, for $i = 1, \ldots, n$; $x_n \in \mathbf{X_n}$ and $x'_n \in \mathbf{X'_n}$,
3) $|g_i - g'_i| \leq e$, for $i = 1, \ldots, n-1$; $g_i \in \mathbf{G_n}$ and $g'_i \in \mathbf{G'_n}$.

Note that checking the valid occurrence of $\mathbf{X'_n}$ may require us to determine all possible gaps between any ordered combination of elements in $\mathbf{SM'_i}$. We use Lemma 1 to expedite this step.

Definition 5.5. (Support) The support of an n-itemset patterns, denoted as sup($\mathbf{P_n}$), is the number of times in which an n-itemset patterns satisfy Definition 9 in **D**.

Definition 5.6. (Frequent Patterns) A n-itemset patterns $\mathbf{P_n}$ is *frequent* if its support is more than or equal to some threshold minimum support (q).

Definition 5.7. (Closed Frequent Pattern) A n-itemset patterns $\mathbf{P_n}$ is a *closed frequent pattern* if there exist no n-itemset pattern $\mathbf{P'_n}$ such that 1) $\mathbf{P'_n} \subset \mathbf{P_n}$ 2)for all transaction $\mathbf{D}$, $\mathbf{P_n} \in \mathbf{D} \rightarrow \mathbf{P'_n} \in \mathbf{D}$.

We can use the naive tree growing procedure to propose a brute-force algorithm for mining frequent patterns. It is easy to see that the brute-force algorithm will return all closed frequent itemsets, as all valid frequent itemsets are enumerated. However it is too inefficient to be executed in practice. It runs in $O(|\mathbf{I}|l^n)$. Below we propose an efficient algorithm to solve this problem.

The idea of the algorithm is to determine all the closed frequent pattern by merging nodes of the tree with all the valid nodes at second level of the tree.

The algorithm is presented in Algorithm 2. In step 1, we initialize the frequent pattern set. A submotif set is obtained in step 2. From this submotif set we initialize level 1 of the tree in step 3. The subsequent tree growing procedure is divided into two parts. First part generates level 2 as shown in step 7-24. We find gaps of all ordered pairs in **D** in step 9-13. Then we obtained all the nodes of level 2 in step 14. Frequent pattern at level 2 is verified in steps 17-24. We use closed frequent pattern at level 2 to generate nodes level 3 onwards as shown in step 26. And closed frequent pattern verification is done in step 29-33.

Definition 5.8. (Generalized A-priori Condition) A n-itemset pattern is frequent only if its parents k-itemset patterns, for all $k = 1, \ldots, n-1$, are frequent.

From this definition we derive the following result:

Lemma 5.2. *Algorithm 2 satisfies the Generalized Apriori Condition (stated in Definition 13).*

Theorem 5.1. *(Correctness of Algorithm 1) Algorithm 1 returns all closed frequent patterns.*

Algorithm 1 Constrained Frequent Pattern Mining

$ConstrainedFP(\mathbf{D}, gap\_threshold(e), q)$

$\mathbf{FP} = \phi$
$\mathbf{I} = \text{Cluster}(\mathbf{D})$
$\mathbf{L_1^0} = \{\mathbf{P_1^1}, \ldots, \mathbf{P_1^{|I|}}\}$, where $\mathbf{P_1^{k_1}} = \{x_{k_1}\}$
$\{\mathbf{G}', \mathbf{I}', \bar{\mathbf{G}}\} = \phi$
$n = 2$
for all $\mathbf{P_{n-1}^{k_{n-1}}} \in \mathbf{L_{n-1}^{k_{n-1}}}$ **do**
  if $(n = 2)$ **then**
    for all $i \in M_i$ **do**
      for all pairs (x_{ij}, x_{ik}) where $x_{ij}, x_{ik} \in \mathbf{X_i'}$ **do**
        compute $g(x_{ij}, x_{ik})$ using Lemma 1
        append $g(x_{ij}, x_{ik})$, $g(x_{ij}, x_{ik}) + e$, and $g(x_{ij}, x_{ik}) - e$ into $\bar{\mathbf{G}}$
      end for
    end for
    $\mathbf{L_2^{k_1}} = \mathbf{P_1^{k_1}} \times \{\bar{\mathbf{G}} \times \mathbf{I}\}$
    for $\mathbf{P_2^{k_2}} \in \mathbf{L_2^{k_1}}$ **do**
      $min\_sup(\mathbf{P_2^{k_2}}) \leftarrow CountSupport(\mathbf{D}, \mathbf{P_2^{k_2}}, e)$
      if $sup(\mathbf{P_2^{k_2}}) \geq q$ **then**
        append $\mathbf{G_2^{k_2}}$ to $\widehat{\mathbf{G}}_{\mathbf{x_{k_2}}}$, where x_{k_2} is last element of $\mathbf{P_2^{k_2}}$
        append $\mathbf{X_2^{k_2}}$ to $\widehat{\mathbf{I}}_{\mathbf{x_{k_2}}}$
        append $\mathbf{P_2^{k_2}}$ to $\mathbf{FP}$
      else
        stop expanding $\mathbf{P_2^{k_2}}$
      end if
    end for
  else
    $\mathbf{L_n^{k_{n-1}}} = \mathbf{P_{n-1}^{k_{n-1}}} \times \{\widehat{\mathbf{G}}_{\mathbf{x_{k_{n-1}}}} \times \widehat{\mathbf{I}}_{\mathbf{x_{k_{n-1}}}}\}$, where $x_{k_{n-1}}$ is last element of $\mathbf{P_{n-1}^{k_{n-1}}}$
    for all $\mathbf{P_n^{k_n}} \in \mathbf{L_n^{k_{n-1}}}$ **do**
      $sup(\mathbf{P_n^{k_n}}) \leftarrow CountSupport(\mathbf{D}, \mathbf{P_n^{k_n}}, e)$
      if $sup(\mathbf{P_n^{k_n}}) \geq q$ **then**
        append $\mathbf{P_n^{k_n}}$ to $\mathbf{FP}$
      else
        stop expanding $\mathbf{P_n^{k_n}}$
      end if
    end for
  end if
  $n \leftarrow n + 1$
end for
RETURN $\mathbf{FP}$

Proof for Lemma 5.2 and correctness analysis of Theorem 5.1 can be found in the full paper version.

5.3. *Significance testing and scoring*

Obtaining frequent motif patterns allows us to minimize spurious instances. But it does not say anything about its biological significance. Weeder [17] introduced an effective method to examine the biological significance by considering input sequences and a background model. This evaluation is tied up with the motif scoring mechanism. We adapt Weeder scoring mechanism and extend it to include motifs with *spacers*.

First of all, we have the table of expected frequency values $\varepsilon(p)$ for oligonucleotides ranging from 6 to 8 bp. They are pre-computed from background sequences from RSAT database [23]. The formula used for computing this frequency is: $\varepsilon(p) = o(p)/T$. Where $o(p)$ is the total number of times p was found in the background sequences, and T is the total number of length m oligos in the background sequences.

For motif longer than 8 bp, that comes with or without spacers the expected frequency is modelled using seventh order Markov chain. Let $p = p_1p_2 \dots p_n$ be an n-mer with n greater than 8. It is computed as follows:

$$\varepsilon(p) = \varepsilon(p_1p_2 \dots p_8) \prod_{i=9}^{n} P(p_i|p_{i-7} \dots p_{i-1}) \tag{1}$$

Let n be a spacer and $\mathbf{n} = \{n_{i1}, \dots, n_{ij}\}$ be set of j number of spacers, at position $i_1, \dots, i_j$ where $j \leq i$. The conditional probability $P(p_i|p_{i-7} \dots p_{i-1})$ of having nucleotide p_i preceded by nucleotides $p_{i-7} \dots p_{i-1}$, is computed by using the expected frequency of 8-mers:

$$P(p_i|p_{i-7}, \dots, p_{i-1}) = \frac{\sum_{n_{i1}} \cdots \sum_{n_{ij}} \varepsilon(p_{i-7}p_{i-6} \dots n_{i1}p_{i-2}n_{ij}p_i)}{\sum_{n_{i1}} \cdots \sum_{n_{ij}} \sum_{\bar{n}_i} \varepsilon(p_{i-7}p_{i-6} \dots p_{i-3}n_{i1}p_{i-1}n_{ij}\bar{n}_i)}$$

Given the motif p by by allowing d substitution, with or without spacers, we use formula (2) to obtain motif score. It consists of *sequence specific score* - $\sigma(p)$ and *background score* - $\beta(p)$. The sequence specific, measures the quality of the motif with the regard to input data set. And background score measures the quality of the motif with regard to background model.

Let $l_1, \dots, l_t$ be the length of input sequence in $\mathbf{S}$. Let d_i be the minimum number of mutations p appears in the i-th sequence. The sequence specific score is given by:

$$\sigma(p) = \sum_i log \frac{1}{\varepsilon(p, d_i).l_i}$$

Furthermore, the background score is given by:

$$\beta(p) = log \frac{o(p,d)}{\varepsilon(p,d).\sum l_i}$$

The final motif score is obtained by adding up the two values: $MotifScore(p) = \sigma(p) + \beta(p)$.

The overall algorithm SPACE can be seen in Algorithm 3. It follows the structure showed in Figure 3.

Algorithm 2 SPACE motif finding algorithm

$FindMotif(l, d, l_s, l_c, e, min\_sup, \mathbf{S})$

Initialize **SEL_MOTIFS** as set of selected motif
Initialize **FP_ALL** as set of all frequent patterns
append $GenCandMotifs(l, d, l_s, l_c, \mathbf{S})$ to **B**
for each $\mathbf{D} \in \mathbf{B}$ **do**
    append ConstrainedFP($\mathbf{D}, e, min\_sup$) to **FP_ALL**
end for
Initialize **FP_INST** as set of instances of the given FP
for each $FP \in$ **FP_ALL do**
    append $FindInstances(FP)$ to **FP_INST**
    $M \leftarrow Find\_Consensus(\mathbf{FP\_INST})$
    $Score \leftarrow Final\_Score(M)$
    append to M,Score to **SEL_MOTIFS**
end for
RETURN Ranked **SEL_MOTIFS**

In the next section we will described the preliminary experimental result.

6. Experimental Results

We perform experiments of Algorithm SPACE on two classes of datasets and report our preliminary results. The first class comprises the 56 datasets that Tompa [21] used for the assessment of motif discovery algorithms. This provides a direct way for comparing the performance of our algorithm with 13 other well known algorithms. However, to further analyze its capabilities we consider 3 synthetic datasets and report performance results. For this case we compare our algorithm with Weeder which was the best performing algorithm on Tompa's assessment experiments.

For performance measurement, we use the same four measures proposed by Tompa [21] namely, *sensitivity (nSN)*, *positive predictive value (nPPV)*, *performance coefficient (nPC)*, and *correlation coefficient (nCC)*. Index n is used to denote that the assessment is done at the nucleotide level instead of site level.

6.1. *Results on Tompa's benchmark data set*

Tompa's benchmark dataset has been constructed based on real transcription factor binding sites drawn from four different organisms [21]. It consists of 56 datasets in total. The number of sequences ranges from 1-35 and the sequence lengths may be up to 3000 bp.

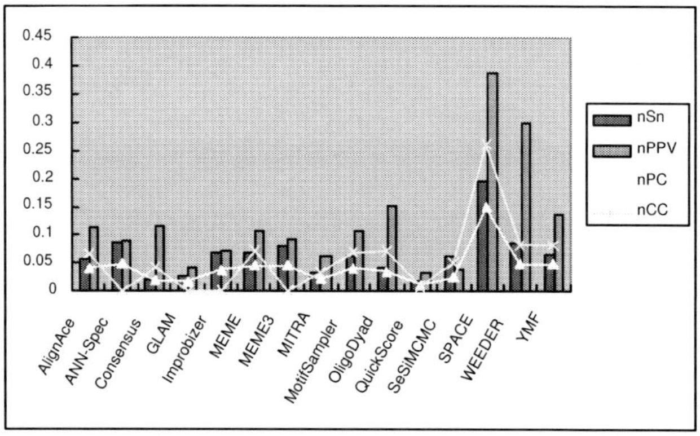

Figure 3. Performance comparison between SPACE algorithm over 13 other motif discovery tools. It represents the total performance value from 4 types of organism in the datasets.

The performance of Algorithm SPACE averaged over all datasets is shown in Figure 3. For this run, we did not use any expert input or per dataset adjustment. It can be observed that our algorithm performs significantly better compared to all other algorithms over all four measures. Compared to the second-best algorithm Weeder, our algorithm achieved more than double sensitivity improvement. We believe this is due to the fact that Algorithm SPACE cannot constrained by the motif length and indeed can find longer motifs. The algorithm also achieved better positive predicted value. As noted in Section 3, this advantage results because of submotif modeling. To illustrate this further, we show the binding sites identified by our algorithm and Weeder on two specific datasets: $hm22m$ and $hm17g$.

Figure 4 shows the motifs output by SPACE and WEEDER on $hm22m$. It can be observed that SPACE found a 12 length motif and WEEDER only an 8 length motif. Figure 5 shows the actual binding sites identified by each algorithm (blue shows the correct binding site and green color shows the binding site found by the algorithm). It can be noted that SPACE does not identify any spurious binding sites.

We also analyze the performance of SPACE across the four organisms. Figure 6 shows the average scores of SPACE for each organism, compared with the best performing algorithm for the respective organism.

On $yeast$, Weeder is the algorithm that did exceptionally well. SPACE achieved only marginal improvements. Improbizer and YMF performed the best on $mouse$ dataset.

Figure 4. Binding sites reported in hm22m (*human*). SPACE with $nSn = 0.26$, $nPPV = 0.62$, $nPC = 0.23$, $nCC = 0.39$ and submotifs: TGGCC, CCACG, CGTGA. Weeder with $nSn = 0.28$, $nPPV = 0.47$, $nPC = 0.21$, $nCC = 0.35$.

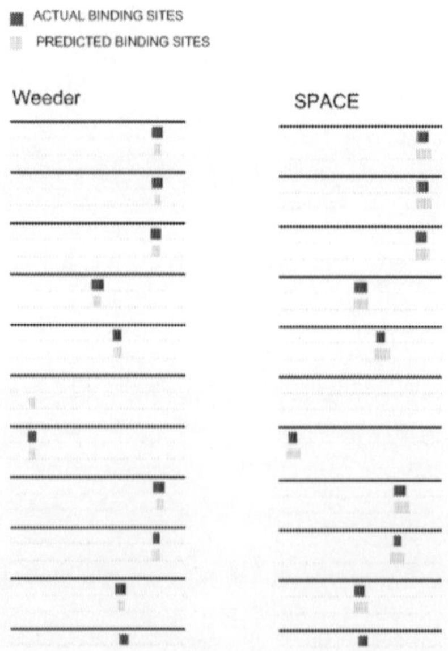

Figure 5. Binding sites reported by SPACE in hm17g (*human*), with $nSn = 0.90$, $nPPV = 0.72$, $nPC = 0.67$ and $nCC = 0.80$. Weeder with $nSn = 0.61$, $nPPV = 0.89$, $nPC = 0.57$ and $nCC = 0.73$.

Though SPACE performed as good as the best in terms of sensitivity, it was significantly better on PPV score. On fly datasets, SeSiMCMC did best in terms of both PPV and sensitivity. SPACE was significantly worse on sensitivity but did well on PPV. However, on $human$ dataset where ANN-Spec was the best-performing algorithm, SPACE achieved significantly better on both sensitivity and PPV. The latter results show evidence that our algorithm can indeed find complex regulatory sites.

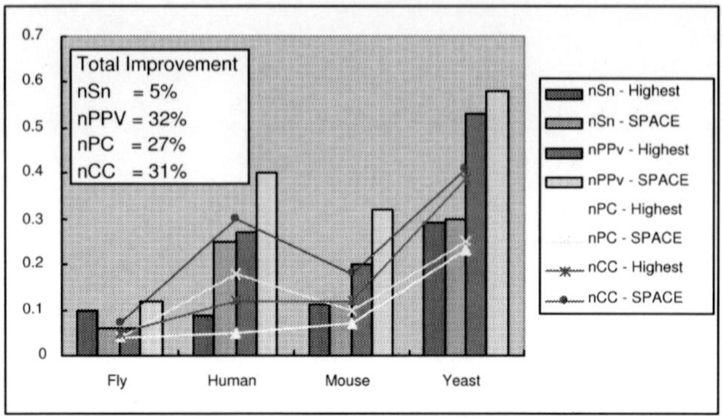

Figure 6. Performance comparison of SPACE and best performing algorithms on 4 types of organisms. Followed by percentage of improvement given by SPACE.

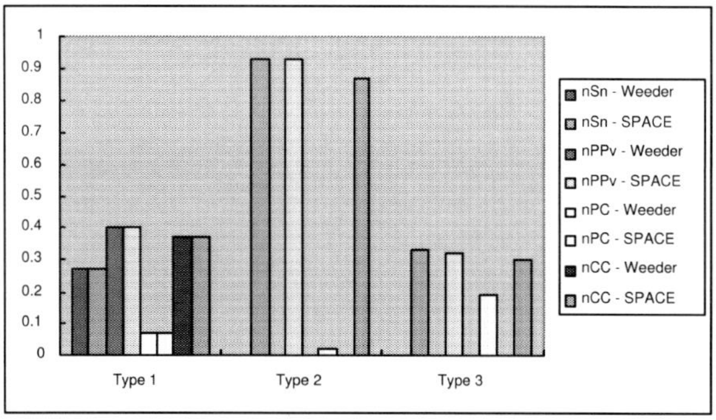

Figure 7. Performance comparison of SPACE and Weeder on 3 synthetic datasets.

6.2. *Results on synthetic data set*

To further analyze the capabilities of SPACE, we we conduct experiments on 3 synthetic datasets created by planting motifs (with and without spacers) on randomly generated sequences. All the 3 datasets contain 10 sequences with the length of each sequence fixed at 300 bp. The motifs implanted in the datasets as as follows: 1) An (8,1)-motif $TGGGTACC$ implanted in 5 out of 10 sequences of 300bp each. 2) A (15,1)-motif $CCTGTNNNAGTTGTC$ containing 2 segments of length 5 and 7 with a spacer of length 3. 3) Same as the motif in (2) but the instances have spacers of varying length 2-4bp.

We compare our algorithm with Weeder which is the best performing algorithm on Tompa's benchmark datasets. We run Weeder in the exhaustive mode (i.e. $large$ mode). The performance results are shown in Figure 7.

For Dataset 1 (simple motif with no spacer), SPACE and Weeder return the same motif $GTACC$ which is a substring of the target motif. On Dataset 2, SPACE returns a spaced motif $CCTGTNNNAGTTG$ whereas Weeder gives a spurious motif $TTAGTATA$. On Dataset 3, SPACE again returns a spaced motif $CTGTANNNNTTGTC$ and Weeder gives $TAATGTTTGT$. On further investigation, we found that Weeder identifies part of the correct motif $GTAGGTG$ albeit at a lower rank.

The results show that SPACE can find spaced motifs with high sensitivity and PPV even when the binding sites contain varying length spacers.

7. Discussion and Conclusions

In this paper we have proposed a new approach for motif-finding through the notion called submotifs. We developed a novel motif-finding algorithm SPACE that offers flexibility in handling variations in spacer length, the number of motif segments and their lengths. SPACE uses a divide-and-conquer approach whereby the target motif is detected by first finding its submotifs and then strategically compositing them using a novel frequent pattern mining compositing them to deduce the target motif. We showed that the algorithm can find multiform TFBS containing spacers.

The existing spaced motif methods generally make several assumptions on spacer length, motif length or number of segments. OligoDyad [22] finds spaced motifs by finding motif pairs with fixed length spacer. Suffix tree approach by Sagot [15] can find multi-segment motifs but the spacer length is required to be fixed. YMF [19] is the algorithm that allows varying length of spacers. However, this is done by enumerating all the possible gap lengths which is computationally expensive and hence is usually limited to finding short motifs. In comparison, SPACE does not assume fixed length spacers and can find longer motifs. This is possible because of the novel constrained frequent pattern mining method employed by SPACE. This method enables structured submotif patterns can be mined from all the submotifs found, without strong assumptions on the spacer/motif length or the number of segments. The fact that the algorithm allows overlapping submotifs also enables the algorithm to find varying length motifs.

This paper reports only preliminary results of our ongoing work. Currently we are optimizing some key components of the algorithm for efficiency, and also testing it on more real biological datasets.

References

1. Ao, W. Gaudet, J., Kent, W. J. et al. (2001), Environmentally induced foregut remodeling by PHA4/FoxA and DAF-12/NHR, *Science*, **305**, 1743–1746.
2. Bailey, T. Elkan, C. (1995), Unsupervised learning of multiple motifs in biopolymers using expectation maximization, *Machine Learning*, **21**, 51–80.
3. Carvalho, A. M., Freitas, A. T., Olivera, A. L. (2003), A Highly Scalable Algorithm for the Extraction of Cis-Regulatory Regions, in *Proceedings of the Third Asia-Pacific Bioinformatics Conference (APBC)*, 273–282.
4. Eisen, M. B. (2005), All motifs are not created equal: structural properties of transcription factor – DNA interaction and the inference of sequence specificity, *Genome Biology*, **6**, P7.
5. Eskin, E. and Pevzner, P. (2002), Finding composite regulatory patterns in DNA sequences, *Bioinformatics*, **18**, S354–S363.
6. Favorov, A. V., Gelfand, M. S, Gerasimova, A. V., Ravcheev, D. A., Mironov, A. A., and Makeev, V. J. (2005), A Gibbs sampler for identification of symmetrically structured, spaced DNA motifs with improved estimation of the signal length, *Bioinformatics*, **21**, 2240–5.
7. Gusfeld, D. (1997), *Algorithm on Strings, Trees, and Sequences: Computer Science, and Computational Biology*, 235–236.
8. Hertz, G. Z. and Stormo, G. D. (1999), Identifying DNA and protein patterns with statistically significant alignments of multiple sequences, *Bioinformatics*, **15**, 563–577.
9. Jiawei, H. and Kamber, M. (2000), Data Mining: Concepts and Techniques.
10. Jensen, K. L., Styczynski, M. P., Rigoutsos, I. and Stephanopoulos, G. (2006), A generic motif discovery algorithm for sequential data, *Bioinformatics*, **22**, 21–28.
11. Down, T. A. and Hubbard, T. J. P. (2005), NestedMICA: sensitive inference of overrepresented motifs in nucleic acid sequence, *Nucleic Acids Research*, **33:5**, 1445–1453.
12. Keich, U. and Pevzner, P. (2002), Finding motifs in the twilight zone, in *Proceedings of the Sixth Annual International Conference on Research in Computational Molecular Biology (RECOMB)*, 195–204.
13. Lawrence, C. and Altschul, S. *et al.* (1993), Detecting subtle sequence signals: a Gibbs sampling strategy for multiple alignment, *Science*, **1993**, 133–154.
14. Lewin, B. (1997), Genes VI, *Oxford University Press*.
15. Marsan, L. and Sagot, M.-F. (2000), Algorithms for Extracting Structured Motifs Using a Suffix

Tree with an Application to Promoter and Regulatory Site Consensus Identification, *Journal of Comp. Biol.*, **7**, 345–360.
16. Owen, G. and Zelent, A. (2000), Origins and evolutionary diversi.cation of nuclear receptor superfamily, *Cell Mol. Life. Sci.*, **57**, 809–827.
17. Pavesi, G., Mauri, G. and Pesole, G. (2001), An algorithm for finding signals of unknown length in DNA sequences, *Bioinformatics*, **17**, S207–S214.
18. Rigoutsos, I. and Floratos, A. (1998), Combinatorial pattern discovery in biological sequences, *Bioinformatics*, **14**, 55–67.
19. Sinha, S. and Tompa, M. (2000), A statistical method for finding transcription factor binding sites, in *Proceedings of the 8th International Conference on Intelligent Systems for Molecular ISMB-00*, 344–354.
20. Thijs, G. *et al.* (2002), A Gibbs sampling method to detect over-represented motifs in the upstream regions of co-expressed genes, *Journal of Computational Biology*, **9**, 447–464.
21. Tompa, M., Li, N., Bailey, T. *et al.* (2005), Assessing computational tools for the discovery of transcription factor binding sites, *Nature Biology*, **23**, 137–144.
22. van Helden, J., Rioas, A. F. and Collado-Vides, J. (2000), Discovering regulatory elements in non-coding sequences by analysis of spaced dyads, *Nucleic Acids Research*, **28**, 1808–1818.
23. van Helden, J. (2003), Regulatory sequence analysis tools, *Nucleic Acids Research*, **31**, 3539–3596.
24. Wijaya, E., Kanagasabai, R., Bramachary, M., Bajic, V. B. and Sung, S.Y. (2005), A Hybrid Algorithm for Motif Discovery from DNA Sequences, *APBC 2006*.
25. Werner, T. (1999), Models for prediction and recognition of eukaryotic promoters, *Mammalian Genome*, **10**, 168–175.
26. Wingender, E., Dietze, P., Karas, H. and Knüppel, R. (1996), TRANSFAC: a database on transcription factors and their DNA binding sites, *Nucleic Acid Research*, **24**, 238–241.
27. Zaki, M. (2001), Spade: An efficient algorithm for mining frequent sequences, *Proceedings of 9th Conf. of Information and Knowledge Management*, 422–429.

REFINING MOTIF FINDERS WITH E-VALUE CALCULATIONS

N. NAGARAJAN, P. NG AND U. KEICH

Department of Computer Science,
Cornell University, Ithaca, NY, USA
E-mail: {niranjan, ppn3, keich}@cs.cornell.edu

Motif finders are an important tool for searching for regulatory elements in DNA. Popular existing programs optimize the entropy score to efficiently search for motifs. While E-values are commonly used for assigning significance to the optimal reported motifs they are not directly optimized for. This raises the question whether optimizing for E-values instead of entropy could improve the finders' ability to detect weak motifs. We first present an efficient algorithm to accurately compute multiple E-values which changes the nature of the above question from a hypothetical to a practical one. Incorporating this method into CONSENSUS- and Gibbs-based finders we then demonstrate on synthetic data that the answer to our question is positive. In particular, E-value based optimizations show significant improvement over existing tools for finding motifs of unknown width.

1. Introduction

The problem of motif-finding can be summarized as scanning a given set of sequences for short, well-conserved ungapped alignments. Most of the interest in this problem comes from its application to identification of transcription factor binding sites, and of *cis*-regulatory elements in general. These in turn are important to the fundamental problem of understanding the regulation of gene expression. This motivated the design of several popular motif-finding tools that search for short sequence motifs given only an input set of sequences (see [11] for a recent comparative review).

Most existing motif finders can be divided into two classes depending on whether they model a motif with a consensus sequence or with a position weight matrix (PWM or profile). Commonly used motif finders that fall in this latter category include MEME [1], CONSENSUS [4] and the various approaches to Gibbs sampling (e.g. [7], [9], [5]). This paper concentrates on improving this class of finders.

Profile-based motif finding algorithms typically try to optimize the entropy score, or information content of the reported alignment which is defined as [10]:[a]

$$I := \sum_{i=1}^{w} \sum_{j=1}^{A} n_{ij} \log \frac{n_{ij}/N}{b_j},$$

[a]Strictly speaking, relative entropy is defined as I/N.

where w is the motif width, n_{ij} denotes the number of occurrences of the jth letter in the ith column of the alignment, b_j is the background frequency of the jth letter[b], N is the number of sequences in the alignment, and A the alphabet size. In order to assign statistical significance to the reported motifs as well as to be able to compare alignments of different widths and depths Hertz and Stormo introduced the notion of a motif E-value. Introduced originally in this context as the "expected frequency" [4], the E-value is the expected number of random alignments of the same dimension that would exhibit an entropy score that is at least as high as the score of the given alignment. When the E-value is high, one can have little confidence in the motif prediction, and conversely when the E-value is low, one can have more confidence in the prediction. It is computed by multiplying the number of possible alignments by the p-value of the alignment. The latter is defined as the probability that a single *given* random alignment would have an entropy score $\geq$ the observed alignment score.

While the E-value is the chosen figure-of-merit for evaluating motifs in popular motif finders such as MEME and CONSENSUS it is not directly optimized for. For example, in MEME E-values are only computed after the EM-algorithm completes its optimization and are only used for significance evaluation and possibly for comparing motifs of different widths. Similarly, when CONSENSUS looks to extend a sub-alignment (matrix) in its greedy search strategy, it chooses the one that optimizes the entropy rather than the E-value[c]. One of the main reasons for this separation between optimization and significance analysis is that E-values are significantly more expensive to compute than entropy scores. Even the relatively fast (and potentially inaccurate [8]) large-deviation method that CONSENSUS employs for computing the E-value can tax an optimization procedure at an unacceptable level.

The discussion above raises two questions:

- Cost aside, can a more direct optimization of the E-value improve our results?
- Can we compute the E-values efficiently so that they can be optimized for?

This paper lays out arguments advocating a positive answer for both questions.

We begin by describing a new technique, memo-sFFT, that allows us to accurately and efficiently compute multiple E-values. We then present the Conspv program that uses the memo-sFFT system to implement a CONSENSUS style motif finder that directly optimizes E-values. The Conspv program generalizes readily to the problem of finding motifs of unknown widths and is functionally equivalent to a combination of CONSENSUS and WCONSENSUS [4]. We show based on experiments on synthetic data that Conspv can significantly improve over WCONSENSUS for finding motifs of unknown widths. As further evidence to the advantage of a

---

[b]Typically estimated from the entire sample.
[c]These two approaches would generally differ if the lengths of the sequences are not identical.

more direct optimization of the E-values, we introduce Gibbspv. This new variant of the Gibbs-sampling algorithm is especially effective when searching for motifs of unknown width by incorporating memo-sFFT to efficiently consider E-values in its optimization procedure. In our experiments on synthetic datasets, Gibbspv clear outperforms other motif finders for finding motifs of unknown width.

It should be noted that GLAM [3] is conceptually quite similar to Gibbspv as both rely on a Gibbs sampling procedure to optimize an overall measurement of statistical significance. However GLAM uses a different significance analysis and as we show below in our tests it is less successful than both Conspv and Gibbspv.

2. Efficiently Computing E-values

In a typical application of CONSENSUS in the experiments described in section 6 about 10^8 alignments are compared. CONSENSUS compares them using entropy scores that can be computed in $O(wn + wA)$ time from scratch, where w is the width of the motif, n is the number of sequences and A is the alphabet size (for our purposes a DNA alphabet of 4 letters). Note that the typical case in CONSENSUS is actually when the score is updated while extending a sub-alignment and this takes $O(w)$ time. In comparison, computing E-values reliably can be many orders of magnitude more expensive if done naively. An efficient algorithm for reliably computing a single p-value (a crucial time-limiting step for computing E-values, see [4]) can typically take $\approx 0.01s$ for the test sets in section 6. This can be prohibitively expensive if incorporated into Conspv (see Table 1).

A partial solution to this problem is to memoize the results. However, we can do even better by relying on algorithms that can compute p-values for a range of scores ([4], [6]). While a single application of these algorithms can be more than 10 times slower, this is compensated for by the fact that they compute a range of p-values that can be stored and reused. We exploit this feature to extend the sFFT algorithm in [6][d] to the memo-sFFT algorithm shown in Figure 1.

In addition we also implemented the following optimizations to memo-sFFT for its use in Conspv and Gibbspv:

- sFFT computes an array p_δ (the pmf of a single column) as the first step in its calculations and this array is independent of the value of w. We utilize this fact and modify sFFT to save and reuse this array across runs.
- The sFFT algorithm requires a lattice size Q (or equivalently a step size δ) that acts as a knob to trade accuracy for speed. We found that setting δ to 0.02 provides good accuracy[e] while being efficient for the experiments in section 6.
- As observed in [6] the sFFT algorithm can typically be used to recover the entire range of p-values (for a given n and w) in a small number (≤ 3)

[d] As shown there, the sFFT algorithm is much more efficient than the numerical method in [4].
[e] Note that the p-value is computed as the geometric mean of the bounds returned by sFFT.

of invocations. In particular, we found that a single well-chosen call to sFFT ($\theta = 1$) can provide a good starting point for memo-sFFT and we implemented this as part of our system.

As can be seen from the results in Table 1, Conspv based on memo-sFFT is indeed much more efficient than a version that computes E-values based on the large-deviation method in CONSENSUS. For the sets described in section 6, we found that less than half a minute is spent in pre-computing p-values in Conspv and the amortized cost of a call to memo-sFFT is essentially that of a table-lookup. The memo-sFFT system therefore opens up the possibility of designing better motif finders that directly optimize the E-value and we present two such algorithms in the next two sections.

```
MEMO-sFFT(n, w, I)
if accuracy[n][w][I] < B
   then (pvalue_sFFT , accuracy_sFFT ) ← sFFT(n, w, I)
        for each I
        do if accuracy[n][w][I] < accuracy_sFFT [I]
              then pvalue[n][w][I] ← pvalue_sFFT [I]
                   accuracy[n][w][I] ← accuracy_sFFT [I]

return pvalue[n][w][I]
```

Figure 1. The memo-sFFT algorithm. Here I is the latticized entropy score [6], B is a desired upper-bound on the relative error (that we set to 10^{-2}) and each entry of array *accuracy* is initialized to a value $\geq B$. Note that we use the term accuracy here to refer to the rigorous bound on the relative roundoff error that can be computed for p-values computed using sFFT [6].

Table 1. The advantage of using memo-sFFT.

Experiment	**memo-sFFT**	**CONSENSUS**
CRP-100	3.0	7.5
CRP-500	3.5	32.7
CRP-1000	4.2	65.1
CRP-5000	9.5	316.6

Note: The columns memo-sFFT and CONSENSUS report the runtime (in seconds) for Conspv implemented with memo-sFFT and the large-deviation method in CONSENSUS respectively, for the various test sets. The CRP-X sets contain 18 sequences of length 108 and X specifies the number of alignments saved by Conspv in its beam search (corresponding to the `-q` option for CONSENSUS).

Table 2. Tests on sequences of varied length.

Experiment	CONSENSUS TPs	Conspv TPs
COMBO1	174	195
COMBO2	146	153
FIFTY1	62	145
FIFTY2	30	41

Note: The values reported here are the number of tests in which the reported motif has a significant overlap with the implanted motif (see section 6 for details) out of a total of 200 tests.

3. Optimizing for E-values - Conspv

The Conspv program in its simplest form adapts the CONSENSUS algorithm with the difference being that it uses E-values rather than entropy scores to compare alignments. More specifically, we implemented a version of the CONSENSUS algorithm under the OOPS model [2] and the `-pr2` option (save the best alignment extension). We also employed the memo-sFFT system described in section 2 to compute E-values. While the cost of computing E-values using this system is essentially a constant, this can still be a significant time penalty for Conspv. We therefore optimized its running time further by not computing the E-value for alignments that have too low an entropy score to be worthy of consideration. This is determined by keeping a lower bound for the entropy score based on alignments that do not make it into the list of best alignments[f].

When run on a set of sequences that have identical lengths the CONSENSUS algorithm (using the entropy score to compare alignments) can be seen as a greedy algorithm to optimize the E-value. However, on a set of sequences of varying lengths this is no longer the case. For such a set, CONSENSUS only optimizes the E-value indirectly. To test if this makes a difference to the performance of CONSENSUS, we compared it to Conspv on some of the test sets described in section 6. As can be seen from the results presented in Table 2, Conspv can significantly improve on the results of CONSENSUS. The improvement is most pronounced on the sets `COMBO1` and `FIFTY1` corresponding to sets where the sequence lengths are more diverged.

A major advantage of Conspv is that it lends itself naturally for searching over multiple motif widths. Since alignments are compared using E-values there is no need for heuristics such as the one used in WCONSENSUS [4]. To exploit this, we implemented a version of Conspv that takes a range of widths to search over as input[g]. The single width version of Conspv is then extended as follows: instead of ranking an alignment by its E-value for a given fixed width, we now rank by the optimal E-value for widths in the given range. A naive alternative approach to this

[f]Recall that CONSENSUS is a beam search algorithm that maintains a list of best alignments seen so far.

[g]A generalization to a set of allowed widths can also be easily implemented.

Table 3. Comparison of CONSENSUS based motif finders.

Experiment	Finders	TPs	Cov	Acc
	WCONSENSUS	52	38	29
COMBO3	WECons	49	40	42
	Conspv	89	76	74
	WCONSENSUS	57	46	43
GAP1	WECons	46	39	38
	Conspv	74	63	60

Note: In the "TPs" column we report the number of tests where there is significant overlap with the implanted motif out of a total of 200 tests. Also, the "Cov" and "Acc" columns report the number of tests in which the overlap is a substantial fraction of the implanted (overlap-coverage) and reported (overlap-accuracy) motifs respectively. For details on the experiments and the overlap scores see section 6.

(that we refer to as WECons) is to run CONSENSUS for each of the widths in the given range and choose the motif with the optimal E-value.

The advantage of Conspv over WECons derives from the fact that the cost of running Conspv for r different widths (where the largest width is w_{max}) is much less than the cost of r runs of CONSENSUS. This is essentially because a majority of the running time of CONSENSUS is spent in evaluating extensions to alignments and this can be done in $O(w_{max})$ time in both CONSENSUS and Conspv. In practice, Conspv is a bit more than twice slower compared to a single run of CONSENSUS. This improved runtime is exploited by Conspv as follows: when searching for the best motif CONSENSUS maintains a list (of size q specified by the user) of the best motifs seen so far. When searching over w different widths, CONSENSUS would maintain w such lists independently. In Conspv a single, much larger list can be maintained for the same total runtime. This enables Conspv to devote more time looking at the promising motifs regardless of their width and thus do a better search of the motif space.

To assess the relative performance of Conspv, WECons and WCONSENSUS[h] we compared them over several synthetic datasets (see section 6 for details). The results were qualitatively similar across the datasets and a couple of them are presented in Table 3. As can be seen in Table 3, Conspv can improve substantially over WCONSENSUS and WECons in finding motifs that overlap the implanted motifs in our datasets. This is also uniformly true across the various overlap scores that we measured and for various thresholds of overlap (as indicated by Figure 2).

4. *E*-value Based Improvements of the Gibbs Sampler

Having established that consideration of E-values can improve the performance of CONSENSUS we next look at the Gibbs sampler. In particular we look at the

[h]Since WCONSENSUS does not allow the user to directly specify a range of widths we instead varied the bias parameter (the `-s` option) over the range $\{0.5, 1, 1.5, 2.0\}$ (as suggested in [4]).

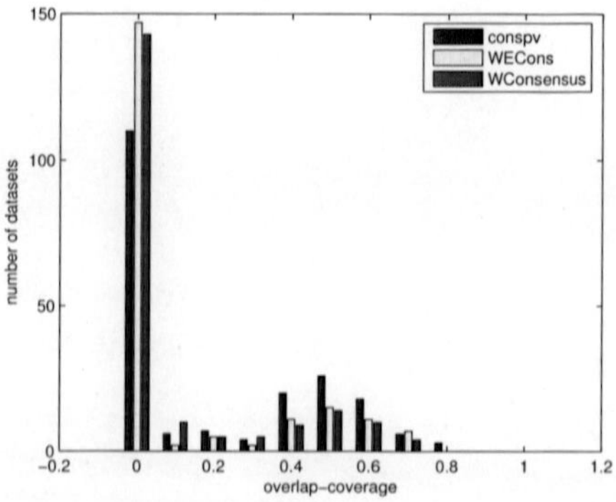

(a)

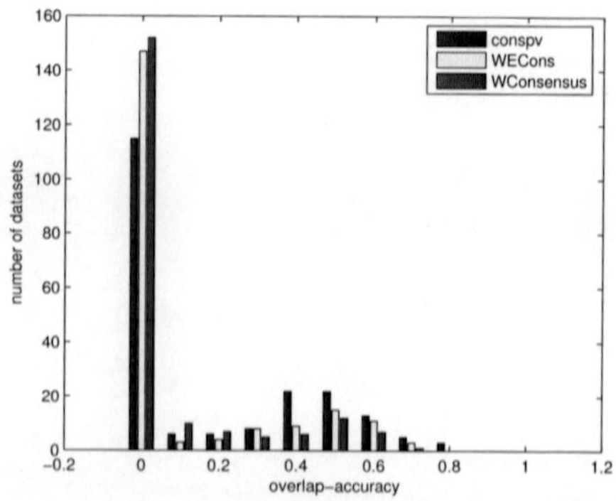

(b)

Figure 2. Histogram of the number of datasets as a function of the overlap score for the COMBO3 experiment and the various motif finders in Table 3.

Table 4. Comparison of Gibbs samplers.

Experiment	Finders	TPs	Cov	Acc
	WGibbs	20	15	17
COMBO3	WEGibbs	125	117	118
	Gibbspv	146	137	136
	WGibbs	17	13	11
GAP1	WEGibbs	77	64	60
	Gibbspv	95	82	79

Note: See Table 3.

problem of unknown motif width. Lawrence et al. [7] considered several criteria for choosing the right width from multiple runs of their sampler, a run for each possible width. The criterion they eventually recommend is termed the "information per parameter" which is the incomplete-data log-probability ratio (22 in [7]) divided by the number of free parameters ($(A-1)w$). Below we refer to this version of Gibbs as WGibbs.

An obvious alternative to WGibbs in the spirit of WECons, which we call WEGibbs, is to choose the run with the width that optimizes the E-value instead of the original information per parameter. Using again the tests described in Section 6 we found that WEGibbs does a significantly better job than WGibbs at detecting the implanted motifs (see Table 4 and Figure 3). The next logical step is to ask whether a Gibbs analogue of Conspv that would more intimately link the E-values to the optimization procedure can further improve these results. To answer this question we designed Gibbspv, a new variant of the Gibbs sampling procedure.

The original Gibbs-sampling motif finder begins each run by picking a random starting position in each sequence in the data set. The algorithm then sequentially applies the following two-step procedure to each of the sample sequences. The pre-

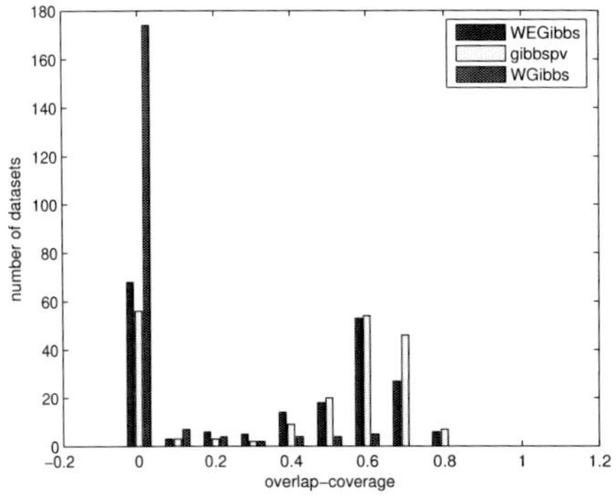

(a)

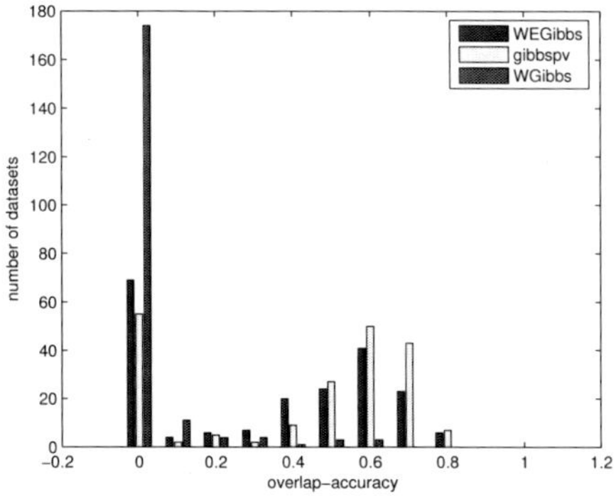

(b)

Figure 3. Histogram of the number of datasets as a function of the overlap score for the `COMBO3` experiment and the various Gibbs samplers in Table 4.

dictive update step computes a motif model Θ based on the current chosen set of starting positions[i]. The sampling step in turn randomly selects new candidate starting positions in the current sequence with probability proportional to the likelihood ratio of the position given the current model Θ. Each iteration of the Gibbs sampler consists of applying the aforementioned two-step procedure once to each of the input sequences.

Gibbspv cycles through a user specified number of iterations (default is `-C`=2) at the end of which five E-values are computed corresponding to the following five alignments[j]:

- The alignment of the currently chosen sites (width w)
- The alignments of first/last $w-1$ columns of the currently chosen sites
- The alignment generated by adding the column to the right/left of the currently chosen sites (width $w+1$)

The algorithm then chooses the alignment with the best (smallest) E-values and continues as before. As in the original Gibbs sampler, if no improvement to the entropy score is detected in a specified number of iterations (`-L`) the program starts a new run.

Table 4 and Figure 3 confirm that by incorporating the E-values into its sampling

[i]The model Θ is inferred from the starting positions by the rule $\Theta_{ij} = \frac{c_{ij}+b_i}{N-1+\sum_j b_j}$, where c_{ij} is the count of letter j in the i-th sequence of the alignment and b_j is an *a priori* chosen pseudocount to avoid 0 probabilities.

[j]Subject to the condition that the considered alignment is well defined and within the specified range of widths.

Table 5. Comparison of Gibbspv and Conspv with MEME and GLAM.

Experiment	Finders	TPs	Cov	Acc
	Gibbspv	146	137	136
COMBO3	Conspv	89	76	74
	GLAM	21	16	14
	MEME	39	28	31
	Gibbspv	95	82	79
GAP1	Conspv	74	63	60
	GLAM	17	7	8
	MEME	41	29	26

Note: See Table 3.

strategy Gibbspv is better at detecting the implanted motifs in our experiments. In addition, as can be seen from the results in Table 5, Gibbspv can be substantially better than existing algorithms such as MEME [1] and GLAM [3] for finding motifs with unknown width.

5. Conclusion

In this paper we have demonstrated the utility of E-value calculations for designing better motif finders. For this purpose, memo-sFFT can serve as an accurate tool for efficiently computing a large number of E-values. In particular, for finding motifs of unknown width, the memo-sFFT based gibbs sampler, Gibbspv can outperform several existing motif finders. Exploring the use of E-values for designing better motif finder for other motif models (such as ZOOPS [2]) can be a fruitful avenue for future research. The programs described in this paper will be available from http://www.cs.cornell.edu/~keich.

6. Methods

To test the various motif finders we constructed synthetic datasets with implanted motifs as follows: independent sequences with the specified lengths were sampled by choosing symbols at random from the four letter DNA alphabet according to a uniform, independent background frequency. A position was chosen uniformly at random from each sequence and an instance of a given profile Θ, generated as described below, was inserted in that position. The profiles used (see Table 6) are represented as a position weight matrix, a $4 \times w$ array of numbers where Θ_{ij} denotes the frequency of letter i in column j in all aligned instances of Θ. Since we wanted to have control over the implanted motifs the instances were essentially generated by permuting the columns of the alignment. Each column of the alignment matched the corresponding column of the profile up to discretizing effects.

For each of the experiments that we conducted, 200 datasets were generated for a given profile. The various motif finders were then run with parameter settings that allowed them to take from 9-10 minutes, to place them on an equal footing. Note that we were unable to do this for MEME as it does not employ any parameters

Table 6. The profiles used in our experiments.

Pos.	COMBO A	COMBO C	COMBO G	COMBO T	FIFTY A	FIFTY C	FIFTY G	FIFTY T	GAP A	GAP C	GAP G	GAP T
1	0.95	0.00	0.00	0.05	0.50	0.00	0.00	0.50	0.70	0.10	0.10	0.10
2	0.00	0.50	0.50	0.00	0.00	0.50	0.50	0.00	0.00	0.70	0.30	0.00
3	0.70	0.10	0.10	0.10	0.50	0.50	0.00	0.00	0.10	0.00	0.90	0.00
4	0.00	0.70	0.30	0.00	0.50	0.00	0.50	0.00	0.10	0.10	0.10	0.70
5	0.50	0.00	0.00	0.50	0.50	0.50	0.00	0.00	0.00	0.70	0.00	0.30
6	0.25	0.25	0.25	0.25	0.00	0.50	0.50	0.00	0.30	0.20	0.30	0.20
7	0.95	0.00	0.00	0.05	0.00	0.50	0.00	0.50	0.25	0.25	0.20	0.30
8	0.25	0.25	0.25	0.25	0.00	0.50	0.00	0.50	0.00	0.50	0.50	0.00
9	0.70	0.10	0.10	0.10	0.50	0.00	0.50	0.00	0.10	0.10	0.70	0.10
10	0.00	0.50	0.00	0.50	0.00	0.50	0.50	0.00	0.00	0.70	0.30	0.00
11	0.00	0.70	0.00	0.30	0.50	0.50	0.00	0.00	0.10	0.10	0.10	0.70
12	0.70	0.10	0.10	0.10	0.00	0.50	0.50	0.00	0.00	0.90	0.10	0.00
13	0.00	0.50	0.50	0.00	0.00	0.50	0.00	0.50	0.30	0.00	0.70	0.00

that allow the control of running time. In all the experiments, MEME ran for much less than 9 minutes. This factor should be taken into account when judging the generally poor performance of MEME compared to the other motif finders. The details for the experiments that we conducted can be found in Table 8. Also, the various profiles used are shown in Table 6.

In general, the length of the sequences and the implanted profiles were chosen such that the motif finders we considered would have a non-trivial percentage of failures (i.e. datasets where they pick motifs with no overlap with the implants). These hard motif finding problems provide good test sets for discriminating between the various motif finders. Finally, an estimate of overlap for each data set and for each motif finder was computed in the following manner: Let a_n be the position of the implanted motif instance in the nth sequence, let $\widehat{a_n}$ be the position of the motif reported by a motif finder and let w and $\widehat{w}$ be the respective widths of the motifs. Then we define the following overlap scores: overlap-coverage = overlap-x$(a, \widehat{a}, w)$, overlap-accuracy = overlap-x$(a, \widehat{a}, \widehat{w})$ and

Table 7. The parameter sets used in our experiments.

Parameter Set	Finder	Parameters
`SINGLE-WIDTH`	CONSENSUS	`-L 13 -c0 -q 4000`
	Conspv	`13 6000`
`MULTI-WIDTH`	WCONSENSUS	`-c0 -q 200`
	WECons	`-c0 -q 120`
	Conspv	`4000`
	GLAM	`-n8000 -r55 -z -a9 -b17`
	MEME	`-mod oops -nmotifs 1 -minw 9` `-maxw 17 -dna -text -maxsize 1000000`
	WGibbs	`-d -n -t80 -L150`
	WEGibbs	`-d -n -t80 -L150`
	Gibbspv	`-t350 -L400`

Note: For the `MULTI-WIDTH` tests the motif-finders were used to search for motifs with widths in the range [9, 17].

Table 8. Experiment details.

Experiment	Profile	Parameter Set	Sequences
COMBO1	COMBO	SINGLE-WIDTH	20 of length 500 & 20 of length 2500
COMBO2	COMBO	SINGLE-WIDTH	20 of length 1000 & 20 of length 2000
FIFTY1	FIFTY	SINGLE-WIDTH	20 of length 500 & 20 of length 2500
FIFTY2	FIFTY	SINGLE-WIDTH	20 of length 1000 & 20 of length 2000
COMBO3	COMBO	MULTI-WIDTH	30 of length 1000
GAP1	GAP	MULTI-WIDTH	30 of length 1000

Note: See Table 6 and 7 for details about the profiles and parameter sets used.

overlap = min{overlap-coverage, overlap-accuracy} where

$$\text{overlap-x}(a, \widehat{a}, x) = \max_{|i|<\frac{x}{2}} \left\{ \frac{x - |i|}{x} \cdot \frac{|\{n : a_n = \widehat{a_n} + i\}|}{N} \right\} \tag{1}$$

and N is the number of sequences in the dataset. To report a significant overlap between the implanted and the reported motif (true positives or TPs) we used a threshold of 0.1 for the overlap score. Also, for overlap-coverage and overlap-accuracy (corresponding to the columns "Cov" and "Acc" in Tables 3, 4 and 5) we used a threshold of 0.3.

Acknowledgements

This research uses computational resources funded by NIH grant 1S10RR020889.

References

1. T.L. Bailey and C. Elkan. Fitting a mixture model by expectation maximization to discover motifs in biopolymers. In *Proceedings of the Second International Conference on Intelligent Systems for Molecular Biology*, pages 28–36, Menlo Park, California, 1994.
2. T.L. Bailey and C. Elkan. The value of prior knowledge in discovering motifs with meme. In *Proceedings of the Third International Conference on Intelligent Systems for Molecular Biology*, pages 21–29, Menlo Park, California, 1995. AAAI Press.
3. Martin C Frith, Ulla Hansen, John L Spouge, and Zhiping Weng. Finding functional sequence elements by multiple local alignment. *Nucleic Acids Res*, 32(1):189–200, 2004.
4. GZ Hertz and GD Stormo. Identifying DNA and protein patterns with statistically significant alignments of multiple sequences. *Bioinformatics*, 15(7-8):563–77, 1999.
5. JD Hughes, PW Estep, S Tavazoie, and GM Church. Computational identification of cis-regulatory elements associated with groups of functionally related genes in Saccharomyces cerevisiae. *J Mol Biol*, 296(5):1205–14, Mar 2000.
6. U. Keich. Efficiently computing the p-value of the entropy score. *J Comput Biol*, 12(4), 2005.
7. CE Lawrence, SF Altschul, MS Boguski, JS Liu, AF Neuwald, and JC Wootton. Detecting subtle sequence signals: a Gibbs sampling strategy for multiple alignment. *Science*, 262(5131):208–14, Oct 1993.
8. Niranjan Nagarajan, Neil Jones, and Uri Keich. Computing the P-value of the infor-

mation content from an alignment of multiple sequences. *Bioinformatics*, 21 Suppl 1(ISMB 2005):i311–i318, Jun 2005.

9. AF Neuwald, JS Liu, and CE Lawrence. Gibbs motif sampling: detection of bacterial outer membrane protein repeats. *Protein Sci*, 4(8):1618–32, Aug 1995.
10. GD Stormo. DNA binding sites: representation and discovery. *Bioinformatics*, 16(1):16–23, Jan 2000.
11. Martin Tompa, Nan Li, Timothy L Bailey, George M Church, Bart De Moor, Eleazar Eskin, Alexander V Favorov, Martin C Frith, Yutao Fu, W James Kent, Vsevolod J Makeev, Andrei A Mironov, William Stafford Noble, Giulio Pavesi, Graziano Pesole, Mireille Rgnier, Nicolas Simonis, Saurabh Sinha, Gert Thijs, Jacques van Helden, Mathias Vandenbogaert, Zhiping Weng, Christopher Workman, Chun Ye, and Zhou Zhu. Assessing computational tools for the discovery of transcription factor binding sites. *Nat Biotechnol*, 23(1):137–44, Jan 2005.

MULTIPLE INDEXING SEQUENCE ALIGNMENT FOR GROUP FEATURE IDENTIFICATION**

WEI-YAO CHOU*, TUN-WEN PAI*$, JIM ZONE-CHANG LAI*, WEN-SHYONG TZOU#

*Department of Computer Science and Engineering & Center for Marine Bioscience and Biotechnology, #Institute of Bioscience and BioTechnology, National Taiwan Ocean University, No. 2,Pei-Ning Rd. Keelung, Taiwan 20224, Republic of China

MARGARET DAH-TSYR CHANG, HAO-TENG CHANG, WEI-YI CHOU, TAN-CHI FAN

Institute of Molecular and Cellular Biology & Department of Life Science, National Tsing Hua University, No. 101 Sec. 2, Kuang Fu Rd. Hsinchu, Taiwan 30013, Republic of China

A novel scheme for combinatorial patterns and exclusive group features identification employing multiple indexing sequence alignment (MISA) based on interval-jumping searching algorithms and hierarchical clustering techniques is proposed in this paper. The interval-jumping searching algorithm transforms sequences into digital number sets in order to find consensus motifs and provides approximate matching results. The searched consensus motifs with tolerant characteristics are labeled and formulated a scoring matrix in order to cluster imported sequences into several subgroups prior to the proposed multiple indexing sequence alignment. To extract distinguishable features among clustered groups, the proposed system performs various combinations of fundamental bitwise operations to obtain their distinctive characteristics. In this paper, MISA has been employed to analyze real biological data and demonstrated to be practical for searching combinatorial patterns for each subgroup and its distinctive features from other subgroups are also identified for further analysis. Comparisons with other existing algorithms are also presented in this paper to demonstrate superior performance of the proposed system.

1. Introduction

Multiple sequence alignment (MSA) is one of the important tasks in the field of computational biology [1]. It reveals conserved regions and evolutionary relationship by direct sequence alignment. Given a set of k sequences ($k \geqq 2$), the MSA problem is to align similar subsequences in the same regions. The optimal alignment of two sequences could be solved in $O(n_1 * n_2)$, where n_1 and n_2 represent the length of two sequences respectively. Unfortunately, when k is strictly greater than two, it becomes an NP-hard problem in an optimal way to match as many residues as possible from the k sequences [2,3]. Therefore, a lot of approximate and heuristic algorithms have been developed to

** This work is supported by Grants NSC 94-2627-B-007-003, NSC 94-3112-B-007-004-Y, VGHUST 94-G2-02-0, and a partial grant from Center for Marine Bioscience and Biotechnology, NTOU, Taiwan, R.O.C..

$ To whom correspondence. E-mail: twp@mail.ntou.edu.tw

overcome such problems intensely. Some algorithms show guaranteed performances and others perform well in practical cases [4,5,6,7].

A natural heuristic for MSA is based on computing optimal pairwise alignments of the k sequences. Several MSA algorithms attempt to combine compatible subsets of optimal pairwise alignment into a multiple alignment. However, they may misalign global multiple residues by matching local pairwise residues. In general, the heuristic alignment performs the process in a one-residue versus one-position manner. Unfortunately, in the evolutionary process, some residues may mutate through transition and transversion which may in turn lead to inaccurate alignment. Although a substitution matrix such as the BLOSUM series matrices can be incorporated to improve the performance, no perfect criterion has yet proved its ability to distinguish "genuine" (biologically relevant) similarities from chance similarities (noise). Failure to align the truly relevant residues may arise from misjudging a noise residue as the target residue. In this paper, we propose an efficient and effective algorithm to solve the above problems. Initially, consensus motifs are identified according to the parameters of pattern lengths, number of tolerant residues and occurrence rates by employing the Ladder-like Interval Jumping Searching Algorithm (LIJSA) [8]. Every subsegment of an input sequence with specified pattern length is scanned, encoded, and allocated in an appropriate numerical interval in sequential order. After searching processes, each consensus motif with various parameter settings possesses a unique indexing number and will be matched in the same interval. Therefore, a sequence in residue representation can be transformed into an indexing notation which is composed of previous labeled consensus segments. The non-matched subsegments are temporarily filtered out for the following alignment procedures. The problem of MSA is converted to a MISA version and the combinatorial pattern extraction becomes the main goal of the proposed system. To approach the aim, we assume that if two transformed indexing sequences contain more consensus motifs simultaneously, then they possess higher similarity. Thus, the system is able to select one sequence as the central sequence that possesses most consensus motifs comparing to all other sequences. Pairwise indexing alignments between the central sequence and all other sequences are performed, and merging operations on pairwise aligned results are implemented to achieve approximate multiple indexing sequence alignment (MISA). The detail procedures will be described in the following.

2. Preliminaries

Let Σ be the set of characters (residues), and $S = \{S_1, S_2, \ldots, S_k\}$ be a set of k sequences with a maximal length of n over Σ. Let $S_i[x \ldots y]$ denote the substring of S_i where $1 \leqq x < y \leqq n$. Let $\Sigma^\#$ be the set of labels of consensus motifs obtained from S by LIJSA with pattern length (m), number of tolerant residues (p_n) and occurrence rate (p_r). $S^\# = \{S^\#_1, S^\#_2, \ldots, S^\#_k\}$ represents a set of transformed indexing sequences with maximal length of $n^\#$ over $\Sigma^\#$, and $S_i^\#[x^\#]$ denotes $S_i[x \ldots x+m-1]$ where $x^\#$ is the index of the new labeled sequence $S_i^\#$, x is the corresponding position of $x^\#$, and m is the specified pattern length.

In general, $n^{\#}$ is much less than n and the differences strongly depend on mutual similarity of sequences. If the substitutable features are considered, $S_i^{\#}$ is then represented as a set of consensus labels due to various encoding values on a specified substring. Each labeling in $\Sigma^{\#}$ represents a pattern with length m and possibly contains up to p_n tolerant residues with occurrence rate greater than or equal to p_r. The pairwise indexing sequence alignment (PISA) of $S^{\#}_1$ and $S^{\#}_2$ performs dynamic programming algorithm by inserting "-" characters and generates two new labeled sequences S'_1 and S'_2 with equal length of n'. A distance function δ $(S'_1[x'], S'_2[y'])$ is defined to measure the relationship between two labels in terms of Hamming distance measurement, where $S'_1[x'], S'_2[y'] \in {}^{\#}\cup \{-\}$, and x' and y' are the indices of the new labeled sequences of S'_i. Thus, the pairwise score of S'_1 and S'_2 is defined as $\Sigma_{1\leq x',y'\leq n'}\delta\,(S'_1[x'], S'_2[y'])$. Once the distance of each indexing sequence pair from the set S' is calculated, an aligned matrix A with k rows and $n'(>= n^{\#})$ columns can be generated by a multi-way merging process. Generally speaking, the designing goal of MISA system is to find the optimal alignment of multiple indexing sequences restrict to the maximal sum of the pairwise alignment.

3. System Architecture

The design of MISA system architecture is motivated by necessity to improve the efficiencies of multiple sequences alignment and to preserve attribute of combinatorial key features in sequential order. Figure 1 shows the configuration of the proposed system. Initially, the system applies LIJSA to search common patterns with different parameter settings, such as numbers of tolerant residues, pattern occurrence rates, and ranges of pattern length. With the searched consensus patterns in the format of encoded numbers, the system replaces these patterns in residues by labeled indices representation according to their relative positions for each sequence. Based on the quantities of local consensus motifs in the indexing sequences, the system provides optional clustering operations by employing hierarchical clustering techniques to classify the imported sequences into several subgroups for extinguishing group feature identification. If there are indeed more than one subgroups embedded in the imported sequence set, the exclusive features of each subgroup can be extracted by performing combination of bitwise operations. On the other

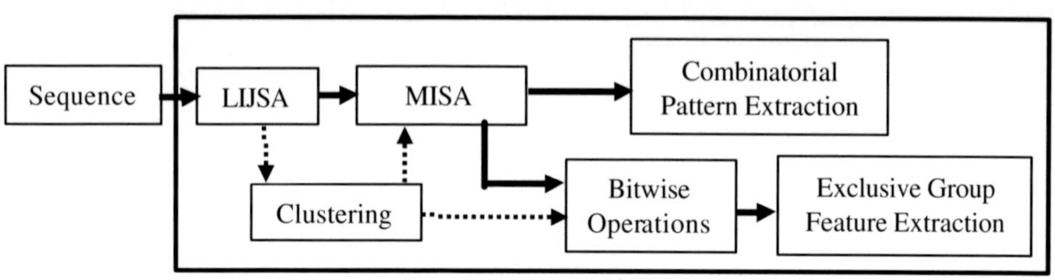

Figure 1. Overall system architecture.

hand, if the imported sequences contain only one family set, then the clustering operation is not required and only one set of combinatorial patterns of such group is identified. The details of each module are described in the following sections.

3.1. *Interval jumping searching algorithm*

To improve efficiency in consensus pattern searching, combination of hashing encoding, quick sorting and interval jumping techniques are applied. Since multiple sequence alignment of bio-sequences from the giant genomic database is usually time consuming, we developed a three-phase methodology in the LIJSA module to search consensus motifs and reduce the time complexity in the comparison. In the first coding phase, bio-sequences are transformed into a numerical space set. The key feature in the encoding phase is that while string sequences are being transformed into numerical sequences, the encoded numbers are with unicity, which can reduce the time complexity comparing with the conventional one-by-one string matching. Subsequently, the quick sort algorithms are employed in the second sorting stage to reorder the encoded data. The main purpose in sorting phase is to match sequentially by the value of numbers to avoid redundant and unsorted sequence searching. In the last searching phase, interval jumping rules are designed to increase efficiencies of numerical comparison. In fact, two interval segmentation techniques, uniform partition and bitwise partition, are applied in order to extend the dissimilarity among encoded numbers in the same interval. The segmenting methodologies are designed according to the length of searching pattern, and the results show that can greatly increase the probability of interval jumping and decrease the cost of matching procedures. To sum up briefly, the LIJSA is a module designed for efficiently discovering consensus motifs in a randomized pattern-matching manner. More details can be found in previous discussions [8].

3.2. *Hierarchical clustering technique*

The purpose of clustering operations is to increase correlation among sequences. With LIJSA processes, sequences possessing high similarities can be grouped together and improve the alignment results, similar to Unweighted Pair-Group Method using Arithmetic averages (UPGMA) clustering techniques that cluster family sequences and exclude outlying sequences [16]. A statistical model concerning the occurrence rates of consensus patterns is applied to build similarity score matrices. Two sequences possessing more consensus patterns are more likely to be clustered into a group based on the assumption of higher order of homologous relationship. The labeled consensus motifs for a grouped subfamily are then aligned according to their indexed positions instead of the original localizations of residues in the primary structure.

3.3. *Multiple indexing sequence alignment*

With the interval-jumping searching processes, $S^{\#}$ is derived from S with customized parameter settings. Whether the optional clustering function is applied in the previous module, for a set of sequences classified as the same group, there will be one sequence defined as the central sequence, $S^{\#}_{c}$, that possesses the largest amount of consensus motifs from all other sequences in that subgroup. Based on the consensus labels found in LIJSA and the selected center indexing sequence, the traditional pairwise alignment by employing dynamic programming is applied on each sequence with respect to the central sequence to carry out all PISAs. Each result of PISA contains two indexed sequences (S'_{ci} and S'_{i}) with appropriate gaps "-" inserted. For clarity, the S'_{ci} is denoted as the PISA result of $S^{\#}_{c}$ between $S^{\#}_{c}$ and $S^{\#}_{i}$, where $i \in [1,k]$. The k pairwise alignments are further combined and transformed into a k-row multiple alignment indexing matrix through a multi-way merging process. The principle "once a gap, always a gap" is obeyed throughout the operations [9], and the steps of multi-way merging algorithm are described as follows.

Algorithm: Multi-way merging
Input: pairwise alignment between k sequences (including $S^{\#}_{c}$ itself)
Output: k-row multiple aligned indexing matrix
Step1: set the initial cursor at the 1st position of each PISA (S'_{ci} and S'_{i})
Step2: scan each cursor position at S'_{ci}
    if any current position appears '-'
      check all other S'_{ci}, if the corresponding position is not '-', then insert '-' into both S'_{ci} and S'_{i}
Step3: shift the cursor to its right position for each sequence
Step4: perform Step2 and Step3 until all the cursor positions are examined
Step5: extract the results of S'_{i} to generate the MISA

3.4. *Group feature identification*

After the MISA is applied on each subfamily, identified consensus and variant residues are respectively denoted by true and false in the output of the system. Binarization and combination of fundamental logical operations among grouped subfamilies are further performed and analyzed. For example, the "AND" operation could be applied to find a global combinatorial patterns of all sequences through individual features of subgroups, and the "XOR" operation is executed to extract local group features which only exists in one specific grouped subfamily and does not belong to local group features of the other subfamily. At the end, the exclusive features of a target subgroup can be clearly retrieved by performing difference operations between the combinatorial patterns of target subgroup and the overlapping combinatorial patterns (by AND operations) from all other subgroups. Several practical examples and comparison with other existing systems are introduced in the experimental results. Therefore, the proposed system can provide an integral solution for discriminative feature identification of each group bases on the combining methodologies of clustering, MISA and bitwise operations, and the algorithms can be summarized step-by-step in the following descriptions.

Algorithm: Multiple Indexing Sequence Alignment for the ith Exclusive Group Feature Identification

Input: k sequences, pattern length, number of tolerant residues and an occurrence rate
Output: combinatorial group feature patterns of a classified group
Step1: perform LIJSA on input sequences
Step2: employ clustering module (optional)
Step3: specify a central sequence $S^{\#}_{c}$ of each cluster
Step4: compute k PISAs with respect to $S^{\#}_{c}$ in each cluster
Step5: execute multi-way merging algorithm in each cluster
Step6: merge overlapped consensus labels to formulate sequential combinatorial patterns of each cluster (CP_i)
Step7: perform bitwise operations for the ith exclusive group features

$$\left(ECP_i = CP_i - \bigcap_{j=1, j\neq i}^{k} CP_j \right)$$

4. Experimental Results

4.1. *Comparison with existing algorithms for combinatorial patterns*

The combinatorial consensus patterns of a set of bio-sequences can be approached by several existing systems. Due to different strategies and goals of developed algorithms, the results of identified combinatorial patterns from various assigned parameters come out remarkably. To demonstrate the validity of the extracted combinatorial patterns from different systems, we select a set of protein sequences which possess known 3D protein structures and compare the goodness (hit numbers and root-mean-square values) by aligning their 3D protein structures constrained to those searched combinatorial patterns. In this example, we select 5 sequences with known protein structures from the RNase A-like superfamily (PDB IDs: 1E21, 1GQV, 1DYT, 1RNF, and 1B1I), and we adopt four systems including MISA, ClustalW (http://www.ebi.ac.uk/clustalw/), MEME (http://meme.sdsc.edu/meme/intro.html), and Gibbs Sampler (http://bayesweb.wadsworth.org/gibbs/gibbs.html) to extract three combinatorial consensus motifs for constrained multiple structural alignment. If the results of structural alignment based on the extracted combinatorial patterns possess higher number of hit residues and lower average root-mean-square deviation values (RMSD), then the combinatorial pattern searching algorithm provides better performance in terms of key feature identification. Here, the extracted combinatorial patterns from different systems are displayed in various formats and colors in Figure 2. The resulting combinatorial patterns identified by the MISA are shown in Figure 2(a), and the system displays the extracted combinatorial features in a very comprehensible format. Interestingly, the key residues of the RNase A-like family, H(12)-K(41)-H(119) (taking 1E21 as an example), are completely identified

Figure 2(a). The combinatorial patterns from MISA system.

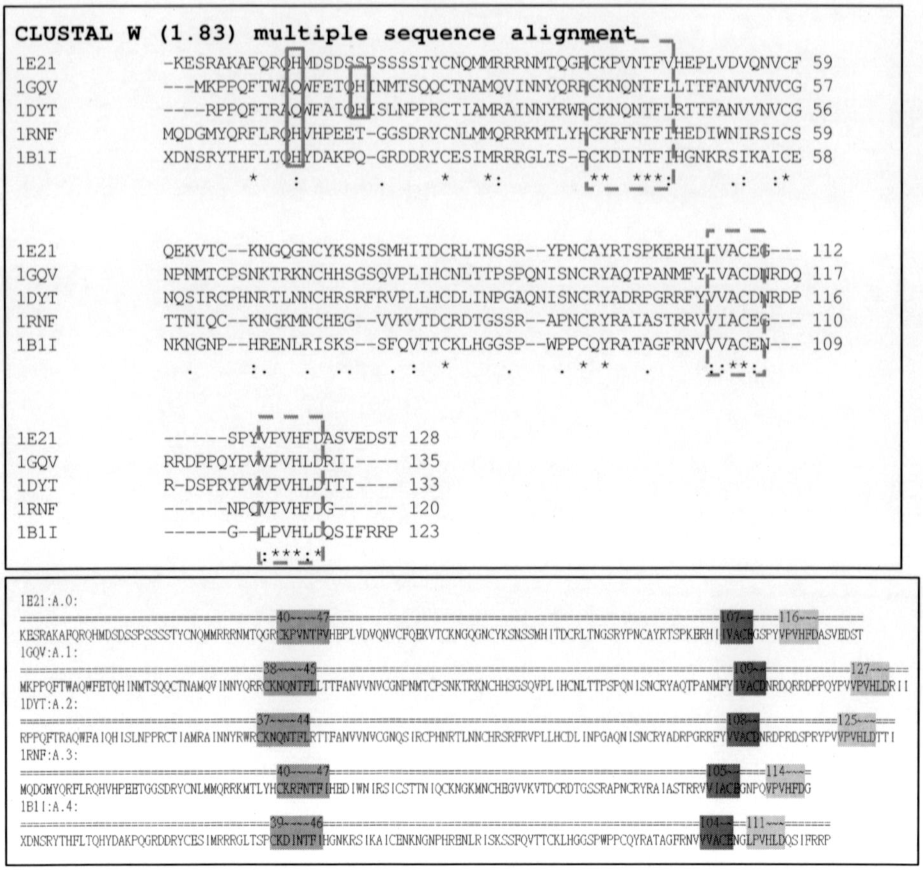

Figure 2(b). The selected combinatorial patterns from ClustalW system.

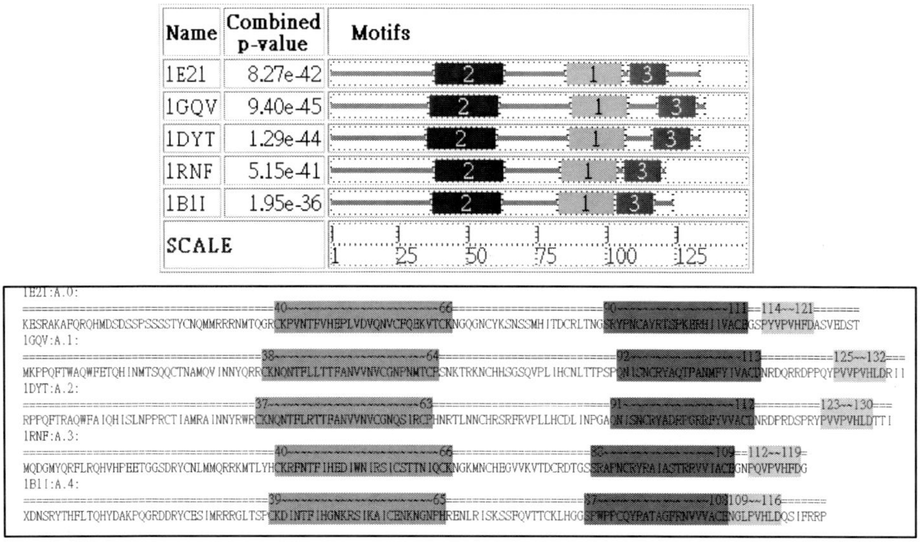

Figure 2(c). The selected combinatorial patterns from MEME system.

1, 1, 1	65 qekvt CKNGQ gncyk	69	1.00 F 1E21:A	Column 1 :	Sequence Number, Site Number
1, 2, 0	107 kerhi IVACE gspyv	111	1.00 F 1E21:A	Column 2 :	Motif type
1, 3, 2	116 egspy VPVHFD asved	121	1.00 F 1E21:A	Column 3 :	Left End Location
2, 1, 1	38 nyqrr CKNQN tfllt	42	1.00 F 1GQV:A	Column 4 :	Motif Element
2, 2, 0	109 anmfy IVACD nrdqr	113	1.00 F 1GQV:A	Column 5 :	Right End Location
2, 3, 2	127 pqypv VPVHLD rii	132	1.00 F 1GQV:A	Column 6 :	Probability of Element
3, 1, 1	37 nyrwr CKNQN tflrt	41	1.00 F 1DYT:A	Column 7 :	Forward Motif (F)
3, 2, 0	108 grrfy VVACD nrdpr	112	1.00 F 1DYT:A		or Reverse Complement (R)
3, 3, 2	125 prypv VPVHLD tti	130	1.00 F 1DYT:A	Column 8 :	Sequence Description from Fasta input
4, 1, 1	65 ttniq CKNGK mnche	69	1.00 F 1RNF:A		
4, 2, 0	105 strrv VIACE gnpqv	109	1.00 F 1RNF:A		
4, 3, 2	114 egnpq VPVHFD g	119	1.00 F 1RNF:A		
5, 1, 1	59 kaice NKNGN phren	63	1.00 F 1B1I:A		
5, 2, 0	104 gfrnv VVACE nglpv	108	1.00 F 1B1I:A		

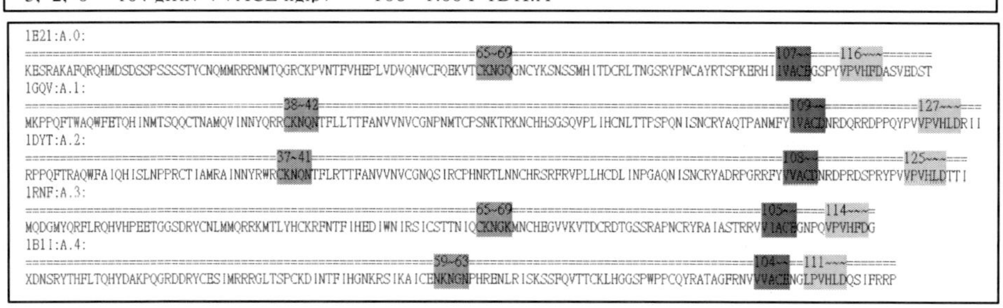

Figure 2(d). The selected combinatorial patterns from Gibbs Sampler system.

Table 1. The comparisons of 3D structural alignment in RMSD values and the number of aligned residues based on the identified combinatorial patterns from MISA, ClustalW, MEME, and Gibbs Sampler.

	MISA	ClustalW	MEME	Gibbs Sampler
Average RMSD value	**1.039139**	1.361162	1.336590	1.406796
Average Number of Aligned Residues	**95.25**	60.50	83.00	38.25

and allocated. In addition to approaching the combinatorial patterns by local segments characteristics, users can employ multiple sequence alignment to observe the sequential combinatorial patterns, and the ClustalW is a well known and popular tool. We apply the ClustalW multiple sequence alignment on these five sequences and the multiple sequences aligned results are shown in Figure 2(b). From the results, we choose the first three longest continuing segments as the key segments for the comparison of key feature verification and 3D structural alignment. These chosen longest segments are framed in dashed-line black boxes and played as the role of geometrical transformation references for constrained 3D structural alignment. Anyway, it can be noticed that the first key residue of "H" is not aligned correctly by ClustalW algorithm due to single residue consideration and annotated in the solid-line black boxes. Therefore, the following 3D structural analysis will not provide a satisfied performance due to the mismatched of key residues. The third illustrated system is the MEME(MAST) which discovers highly conserved regions in groups. The results of MEME system is shown in Figure 2(c), the first three segments identified by MEME are shown sequentially. However, there are only two key residues of RNase A-like family allocated and coincided with MISA's results. The last results are performed by Gibbs Sampler algorithm and shown in Figure 2(d). It is obvious that the first three segments identified by Gibbs Sampler algorithm approached only one correct key feature of RNase family, and the first segment are not matched at the right position neither. In order to prove the importance and correctness of combinatorial feature segments, we utilize the extracted combinatorial patterns from various systems as constrained conditions for multiple structural alignment. The results are shown in Table 1 and it indeed reveals that the combinatorial patterns identified by our proposed systems providing better performance in terms of both the number of alignment residues and the average RMSDs. In this case, the combinatorial key features selected from MISA system result in the lowest average RMSD value (1.039139) and the highest average number of hit residues (95.25) than other three systems. The 3D multiple structural alignment is performed by straight forward translation and rotation operations on structures with respect to the three key residues from three identified combinatorial patterns.

4.2. *Transcription elements prediction*

Another example to verify the practical applications of MISA, the upstream 1.2 kb regions of several primate genes of interest are amplified and sequenced by polymerase chain reactions (PCR). The obtained sequence sets are analyzed by MISA to identify the group features that may serve as the key transcription elements for the regulation of gene expression. As shown in Figure 3, thirty one discrete feature segments in the upstream 1138 bp sequences of *map1b* can be identified and labeled in color-coded bars, the grey grid bars represent regions lacking of common features, and their heights represent the levels of residue variation. Interestingly, a continuously conserved region with 400 nucleotides in length (-400 to -1) containing the experimentally verified *cis*-acting elements CRE (-346/-339), neuronal motif (-323/-296), TCC repeat motif (-279/-257), and Sp1 (-106/-98) critical for *map1b* gene regulation can be rapidly identified, indicating that these sequence motifs are so crucial for regulation of the downstream genes such that they are conserved in all primate species tested [10].

The upstream regions of primate *hspa2* genes are also PCR amplified sequenced and analyzed by MISA. As shown in Figure 4 overall 28 discrete feature segments in the upstream 993 bp sequences of primate *hspa2* are obtained and labeled in color-coded bars. The grey grid bars represent regions lacking of common features, and their heights represent the levels of residue variation. Different from previous results, the conserved sequence motifs appear to be located in the -300 to -800 regions where no regulatory element has been revealed.

Since no transcription element has ever been reported for this particular gene set, the conserved sequence motifs are further analyzed employing ConSite database (http://www.phylofoot.org/consite) and displayed using JASPAR logo as shown in Table 2 [11]. The putative transcription factor binding sites FREAC-3, SPI-1, RORalpha-1, GATA-3, and FREAC-7, are thus marked in Figure 4. These results provide useful information for experimental verification employing genetic manipulation.

4.3. *Exclusive group feature identification*

To demonstrate the function of extraction of exclusive group features, the protein sequences of primate RNase2 and RNase3, a pair of gene duplication product after the

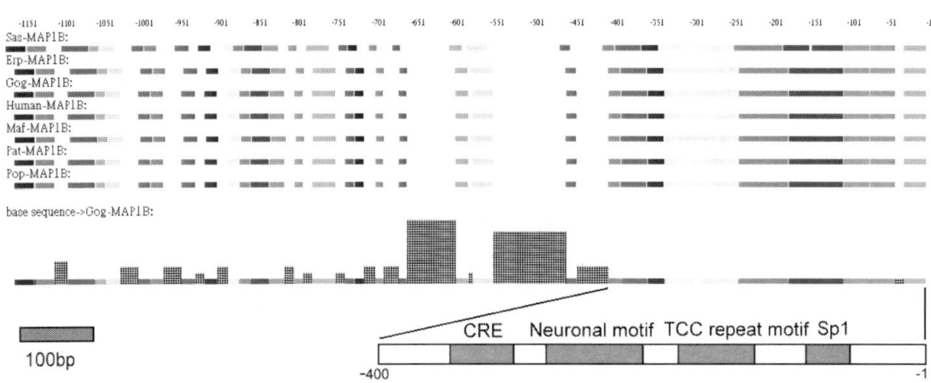

Figure 3. Multiple indexing sequence alignment of primate *map1b* upstream sequences. The labeled positions are relative to the human *map1b* promoter region (-1138 to -1)

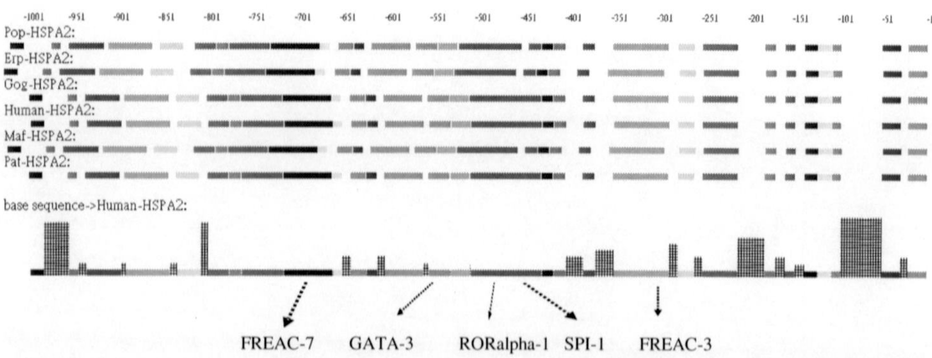

Figure 4. Multiple indexing sequence alignment of primate *hspa2* upstream sequences. Positions are relative to the human *hspa2* promoter region (-993 to -1)

Table 2. Putative transcription factor binding sites on *hspa2* upstream sequences (the logo represents position weight matrices)

Transcription factor	Sequence	Site (Human)	Strand	LOGO
FREAC-3	GCCATGTA	-324/-314	+	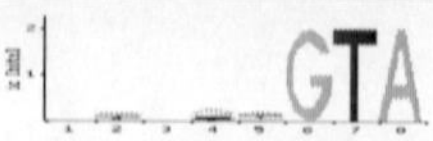
SPI-1	ATTCCC	-455/-450	-	
RORalpha-1	TCACCTTGAA	-479/-470	-	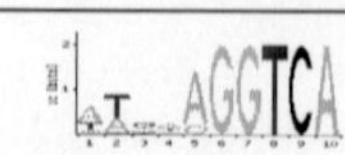
GATA-3	AGAGAG	-580/-575	+	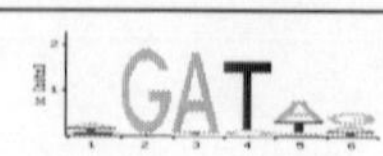
FREAC-7	TTGACATA	-686/-679	+	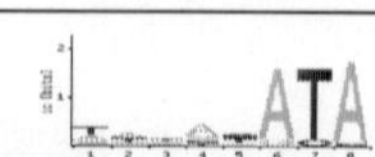

separation of Old World Monkey and New World Monkey, are analyzed and discussed. It is found that very high sequence homology existed in these two RNase data sets. Under the proposed MISA system and bitwise operations, the output for group features is shown in Figure 5(A). The continuous color-coded bars indicate the combinatorial consensus patterns with extremely high similarity located in the RNase2s from three primate species, the RNase3s from the same primate species, as well as the global common features from all six sequences of RNase2+RNase3. The individual group features of RNase2 and RNase3 are identified by applying "XOR" operations to subtract the global consensus patterns of RNase2+RNase3 from the respective consensus patterns of RNase2 and RNase3. The unique features of each subgroup are shown in the red color-coded bars, and

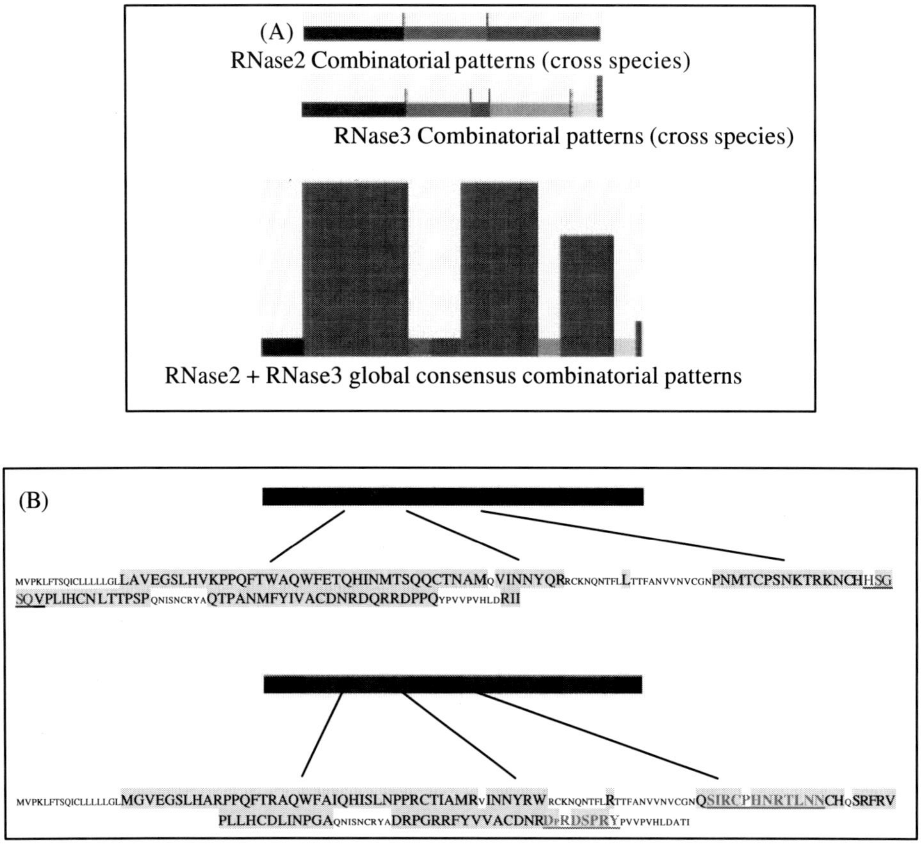

Figure 5. MISA and group features of primate RNase2 and RNase3. (A) Combinatorial features of primate RNase2, RNase3, and RNase2/RNase3. (B) Specific subgroup features of primate RNase2 and RNase3.

corresponding residues are stretched out and highlighted in cyan color in Figure 5B, they apparently matche the grey grid bars in Figure 5A. Interestingly, the specific group features of RNase2 and RNase3 result in identification of several epitope regions where antibodies can specifically distinguish RNase2 and RNase3 by recognizing specific segments labeled in purple, pink and orange [12]. Taken together, we have demonstrated that the proposed novel algorithms can be applied to analyze multiple nucleic acid and protein sequences. The correlations between the sequence motifs identified by computational methods with specific biological functions have been further achieved.

5. Conclusions

During evolutionary processes, a nucleotide residue may mutate into another one to cause misalignment. Even if a substitution matrix can be applied to improve the comparison, a one-residue versus one-position alignment may still result in mismatched alignment. In this paper, we propose a novel scheme for searching combinatorial patterns which is based on local segment characteristics and bottom up strategy. The labeled components with tolerant characteristics are employed as basic elements for performing multiple indexing sequence alignments, and satisfactory results can be approached in an efficient and effective manner.

In the future application, such as constrained multiple sequence alignment [13, 14], specified patterns can be incorporated into our multiple indexing sequence alignment. On the other hand, k-different approximate string matching approaches [15] and substitution concept can be incorporated in our labeling mechanism to make tolerance more statistically enabling and significant.

References

1. S. Chan, A. Wong and D. Chiu. A survey of multiple sequence comparison methods. *B. Math. Biol.*, 54:563–598, 1992.
2. P. Bonizzoni, and G. Vedova. The complexity of multiple sequence alignment with SP-score that is a metric. *Theoretical Computer Science*, 259:63–79, 2001.
3. L. Wang and T. Jiang. On the complexity of multiple sequence alignment. *In Journal of Computational Biology*, 1:337–348, 1994.
4. D. Gusfield. Efficient methods for multiple sequence alignment with guaranteed error bounds. *B. Math. Biol.*, 30(30):141–154, 1993.
5. V Bafna., E. Lawler and P. Pevzner. Approximation algorithms for multiple sequence alignment. *In Theoretical Computer Science*, 182:233–244, 1997.
6. P.A. Pevzner. Multiple alignment, communication cost, and graph matching. *Applied Mathematics*, 52:1763–1779, 1992.
7. C. Notredame, D. Higgins and J. Heringa. T-coffee: A novel method for multiple sequence alignments. *In Journal of Molecular Biology*, 302:205–217, 2000
8. J. Chu, W. Chang, T. PAI, D. Chang, H. Tai, H. Chang. Approximate Matching Using Interval Jumping Searching Algorithms for Sequences. *ICS2004*, 994–999, 2004.
9. D. Feng and R. Doolittle. Progressive sequence alignment as a prerequisite to correct phylogenetic trees. *Journal of Molecular Evolution*, 25:315–360, 1987.
10. D. Liu and I. Fischer. Structural analysis of the proximal region of the microtubule-associated protein 1B promoter. *J. Neurochem*, 69:910–919, 1997.
11. A. Sandelin, W. Wasserman, and B. Lenhard. ConSite: web-based prediction of regulatory elements using cross-species comparison. *Nucleic Acids Res.*, 32:249–252, 2004.
12. H. Chang, T. Pai, T. Fan, B. Su, P. Wu, C. Tang, C. Chang, S. Liu, and D. Chang. Reinforced merging methodology for mapping unique peptide motifs in members of protein families. *BMC Bioinformatics*, 7:38, 2006.
13. Y. Chin, N. Ho, T. Lam, W. Wong, M. Chan. Efficient Constrained Multiple Sequence Alignment with Performance Guarantee. *Proceedings of the IEEE Computer Society Conference on Bioinformatics*, 337, 2003.

14. C. Tang, C. Lu, D. Chang, Y. Tsai, Y. Sun, K. Chao, J. Chang, Y. Chiou, C. Wu, H. Chang, and W. Chou. Constrained multiple sequence alignment tool development and its application to rnase family alignment. *In Proceedings of the First IEEE Computer Society Bioinformatics Conference*, 127–137, 2002.
15. G. Landau and U. Vishkin. Fast String Matching with k Differences. *Journal of Computer and System Sciences*, 37(1):63–78, 1988.
16. P. Sneath and R. Sokal. Numerical Taxonomy. W.H. *Freeman and Company*, San Francisco, 1973.

IMPROVING THE ACCURACY OF SIGNAL TRANSDUCTION PATHWAY CONSTRUCTION USING LEVEL-2 NEIGHBOURS

THOMAS K.F. WONG, S.M. YIU, T.W. LAM AND SIMON C.K. WONG

Department of Computer Science
The University of Hong Kong
Hong Kong
E-mail: {*kfwong, smyiu, twlam, ckwong3*}*@cs.hku.hk*

In this paper, we consider the problem of reconstructing a pathway for a given set of proteins based on available genomics and proteomics information such as gene expression data. In all previous approaches, the scoring function for a candidate pathway usually only depends on adjacent proteins in the pathway. We propose to also consider proteins that are of distance two in the pathway (we call them Level-2 neighbours). We derive a scoring function based on both adjacent proteins and Level-2 neighbours in the pathway and show that our scoring function can increase the accuracy of the predicted pathways through a set of experiments. The problem of computing the pathway with optimal score, in general, is NP-hard. We thus extend a randomized algorithm to make it work on our scoring function to compute the optimal pathway with high probability.

1. Introduction

While more and more DNA genomic sequences of organisms are available, the focus of modern biology gradually shifted to functional genomics. Studying the pathways of the cellular activities is one of the major topics in this area. In this paper, we focus on signal transduction pathways. Signal transduction is an important biological process by which cells convert an external signal into a response. In this process, through a series of protein interactions and activations, the signal is transduced. A major challenge to understand the process is to determine the order of these protein interactions and activations. Roughly speaking, we are given a set of proteins, the problem is to reconstruct the pathways, that is, determine the order of proteins in each pathway.

One can use laboratory-based techniques (for example, genetic epistasis analysis) to determine the ordering of proteins in the pathways. However, the experiments are both time-consuming and expensive. To reduce the cost, computational techniques were recently introduced to predict possible pathways based on available genomics and proteomics information [6, 1, 3] such as gene expression profiles and the database of interacting proteins (DIP [7]). All these techniques are based on the following rationale. If two proteins are involved in the same signal transduction pathway and are adjacent to each other, there should be some available information showing that they are somehow related. For examples, their corresponding gene expression profiles should be very similar. Or they may have the same protein domains. Or they may have interactions which are documented in the DIP. Then, for each generated pathway, a score based on these existing information can be calculated to rank whether this pathway is likely to be a true pathway or not.

Preliminary results on yeast from these research projects show that the direction is promising although most of the ideas are still in the initial stage. In [1], the authors proposed to predict the signal transduction pathways using the information of protein domains and protein interaction database. However, not all proteins have a defined domain composition. On the other hand, information for gene expression profiles is more available. In particular, gene expression profiles for all genes in yeast under more than 300 different conditions are available [4, 2] for download[a]. Thus in [6], the authors proposed to use gene expression data instead of the protein domain information. They first construct possible pathways based on the information in DIP, then rank the pathways using the gene expression profiles. Their approach has a limitation that if DIP does not contain the information for the adjacent proteins in the true pathway, that pathway will not be generated and thus will not be considered in the subsequent scoring phase. From their experiments, one can notice that some of the correct pathways will never be reported due to the above limitation.

To rectify this deficiency, instead of using the DIP information to construct possible pathways, Liu and Zhao [3] considered all possible pathways. However, the number of possible pathways may be tremendous. In practice, researchers may have information about what proteins are actually involved in the process together with which proteins start and end the process. So, they considered a restricted version of the problem and only reconstruct one pathway at a time. They assume that all the pathway components (the proteins) are known and we are given the proteins that start and end the process. The problem is to order the proteins to form the pathway. They examined all permuted orderings of the pathway components and make use of the information in DIP and gene expression profiles to rank each possible pathway.

They calculate the score for each pathway as follows. They consider every pair of adjacent proteins in the pathway. For each pair, they calculate two scores. The first score is the correlation coefficient of their corresponding gene expression profiles. Based on whether this pair of proteins have interaction (using the information of DIP), another score is computed. The final integrated score for this pair of adjacent proteins is the sum of the normalized scores of the two scores. The sum of the integrated scores of all adjacent proteins is the score for the pathway. They have evaluated their approach using several yeast MAPK pathways. Some of these true pathways are ranked very high which show that using DIP and gene expression data is a promising direction. To further investigate the effectiveness of their approach, we have applied their approach (we call it LZ Approach) to other pathways related to the MAPK processes, however, the result is not very satisfactory. About half of the true pathways are ranked below 50 and some of them are with rank more than 100 (see Table 1 for the pathways we used in our experiments and the first column "LZ Approach" in Table 2 for the ranking of these pathways using their approach).

Another issue of their approach is that if the pathway is long, the number of possible

[a]http://pubmed-central.nih.gov/articlerender.fcgi?artid=526379#N0x860db48.0x903f358

Table 1. The MAPK Pathways

Index	MAPK Pathway	Total No. of Permutations
Functionality: Pheromone		
1	STE2 → STE4 → STE18 → DC42 → STE20 → STE11 → STE7 → FUS3 → DIG1 → DIG2 → STE12	362880
2	STE3 → STE4 → STE18 → DC42 → STE20 → STE11 → STE7 → FUS3 → DIG1 → DIG2 → STE12	362880
Functionality: Hypotonic Shock		
3	MID2 → RHO1 → PKC1 → BCK1 → MKK1 → MKK2 → SLT2 → RLM1	720
4	MID2 → RHO1 → PKC1 → BCK1 → MKK1 → MKK2 → SLT2 → SWI4	720
5	MID2 → RHO1 → PKC1 → BCK1 → MKK1 → MKK2 → SLT2 → SWI6	720
6	WSC1 → RHO1 → PKC1 → BCK1 → MKK1 → MKK2 → SLT2 → RLM1	720
7	WSC1 → RHO1 → PKC1 → BCK1 → MKK1 → MKK2 → SLT2 → SWI4	720
8	WSC1 → RHO1 → PKC1 → BCK1 → MKK1 → MKK2 → SLT2 → SWI6	720
9	WSC2 → RHO1 → PKC1 → BCK1 → MKK1 → MKK2 → SLT2 → RLM1	720
10	WSC2 → RHO1 → PKC1 → BCK1 → MKK1 → MKK2 → SLT2 → SWI4	720
11	WSC2 → RHO1 → PKC1 → BCK1 → MKK1 → MKK2 → SLT2 → SWI6	720
12	WSC3 → RHO1 → PKC1 → BCK1 → MKK1 → MKK2 → SLT2 → RLM1	720
13	WSC3 → RHO1 → PKC1 → BCK1 → MKK1 → MKK2 → SLT2 → SWI4	720
14	WSC3 → RHO1 → PKC1 → BCK1 → MKK1 → MKK2 → SLT2 → SWI6	720
Functionality: High Osmolarity		
15	SHO1 → STE20 → STE11 → PBS2 → HOG1 → MSN2	24
16	SHO1 → STE20 → STE11 → PBS2 → HOG1 → MSN4	24
17	SLN1 → YPD1 → SSK1 → SSK2 → PBS2 → HOG1 → MSN2	120
18	SLN1 → YPD1 → SSK1 → SSK2 → PBS2 → HOG1 → MSN4	120
Functionality: Starvation		
19	RAS2 → CDC42 → STE20 → STE11 → STE7 → KSS1 → DIG1 → DIG2 → STE12	5040
20	SHO1 → CDC42 → STE20 → STE11 → STE7 → KSS1 → DIG1 → DIG2 → STE12	5040

candidate pathways may be huge. Also, the assumption of knowing all pathway components may not be realistic. In practice, usually we are given a set of proteins in which quite a number of them (called noise) are not involved in the pathway. In fact, the amount of noise may be more than the number of proteins actually involved in the pathway. In any case, the number of possible pathways increases if the number of input proteins increases. For a pathway of length 11, we estimate that the LZ approach may take more than 7 days to check all possible ordering of the proteins. In fact, given a set of proteins, the proteins that start and end the process, the length of pathway, and a scoring function, the problem of computing the pathway with optimal score is NP-hard [5].

Our Contributions:

In this paper, similar to [3], we consider the pathway reconstruction problem based on information in DIP and the gene expression data. We focus on the problem of constructing one pathway and we study both the restricted version of the problem (only the proteins that are involved in the pathway are given as input) and the version with noise in the input. Our findings include the followings.

- In [3], when calculating the score of a pathway, we only consider adjacent proteins (we call this Level-1 neighbours) in the pathway. However, a preliminary study on the DIP database shows that the database also captures the interaction information for proteins that are Level-2 neighbours (that is, they are of distance two in the pathway). So, we derived a scoring function based on both Level-1 and Level-2 neighbours.

Table 2. Ranking of the True Pathways using LZ Approach and Our Approach

Index	LZ Approach	10% Gene Expression Data Removed	
		LZ Approach	Our Approach
1	3	2	2
2	59	14	11
3	18	15	5
4	66	20	6
5	104	14	6
6	57	42	5
7	234	62	6
8	341	46	6
9	21	14	5
10	83	19	6
11	145	13	6
12	13	14	5
13	50	23	6
14	79	12	6
15	7	6	7
16	13	8	10
17	2	1	1
18	2	1	1
19	139	3	1
20	20	2	1

- When using the gene expression data, there may be noise in the data, we propose a simple heuristics to remove some of the noise before it is used in the prediction process.
- To handle large input, in [5], the authors provided a randomized algorithm to solve the problem with high probability. We extend their algorithm to handle our scoring function.

We have evaluated our solution using the 20 pathways given in Table 1. Experimental results show that our solution is effective. In particular, (1) we show that the cleansing of the gene expression data before we use it in the reconstruction process is effective. There is a significant improvement in the results even for LZ approach. The ranking of the true pathways have been raised to within 50 for all cases except one which has a rank of 62 (see Table 2). (2) Our scoring function is more effective than the one used in LZ approach. For the restricted version of the problem, all the true pathways have ranks within 10 except one that has a rank of 11. For the version with noise in the input, the ranking of the true pathways also shows a significant improvement over the LZ approach.

2. The Problem

The problems we study in the paper are as follow. We are given a set of proteins which are known to be involved in the same pathway. We are also given the proteins that start and end the pathway. We assume that for these proteins, the corresponding gene expression data under different conditions are given. We also have the information in DIP which captures whether two proteins have interaction or not. The problem is to reconstruct the pathway (order the proteins). The output of the solution is expected to be a list of possible pathways with ranks. The pathway with a higher rank should be more likely to be the true pathway.

We also consider a variation of the problem that the input proteins contain noise. In other words, in addition to the proteins that are actually involved in the pathway, the input proteins will contain other proteins that are not in the pathway. In this version of the problem, we assume that the length of the pathway is given as a parameter.

3. Our Approach

3.1. *The scoring function*

It is generally believed that if two proteins are adjacent in the same pathway, their corresponding gene expression profiles should be similar. Intuitively, this observation should also be valid for proteins which are close in the pathway. We call two proteins in the same pathway Level-2 neighbours if they are of distance two in the pathway. Level-2 neighbours should also have similar gene expression profiles.

The DIP database captures whether two proteins have interaction or not. From a preliminary study of the DIP database, we observed that if DIP shows a relationship between two proteins, then they have a high chance to be adjacent in the same pathway. On the other hand, for a certain percentage of Level-2 neighbours, DIP also declares that they have interaction. This may be due to the fact that the interaction between proteins may occur very fast, so the researchers may assume that the two proteins interact and miss the intermediate protein or in fact, the three proteins interact together instantly. We have verified this observation by checking the statistics in the DIP database using the MAPK pathways. Among the interactions of the proteins involved in these pathways, we found that about 80% of them are between adjacent proteins while 16% of them are between Level-2 neighbours.

Based on the above observations, we derive a scoring function based on both adjacent proteins and Level-2 neighbours in the pathway. For any two proteins x and y, let $\text{Dip}(x, y)$ be the probability that x and y are related in the same pathway. If the DIP database indicates that they interact, then we set $\text{Dip}(x, y) = \text{tp}$ (the true positive probability). Otherwise, we set $\text{Dip}(x, y) = \text{fn}$ (the false negative probability). Let $\text{Corr}(x, y)$ be the absolute value of the correlation coefficient of the gene expression profiles of x and y. Given a pathway $P = x_1, x_2, \ldots, x_k$ with length k, we calculate the score of P as

$$\sum_{i=1}^{k-1} \text{Corr}(x_i, x_{i+1}) \times \text{Dip}(x_i, x_{i+1}) + \sum_{i=1}^{k-2} \text{Corr}(x_i, x_{i+2}) \times [\text{Dip}(x_i, x_{i+2})]^2.$$

The first summation considers the adjacent proteins while the second summation considers all the Level-2 neighbours. We remark that we have tried to omit the second summation in the scoring function, that is, we do not consider Level-2 neighbours. However, our experiments show that there is a significant drop in accuracy if we only consider Level-1 neighbours (adjacent proteins). In our experiments, we set $\text{tp} = 0.8$ and $\text{fn} = 0.1$.

3.2. *The randomized algorithm*

Recall that in real situation, we seldom know exactly what proteins are involved in a pathway. Usually, we are given a set of proteins which contain quite a number of proteins that are not involved in the pathway. In that case, the number of permutations may increase with the number of proteins given in the input. Or if the length of the pathway is long, the number of candidate pathways may be huge. It is NP-hard to compute the pathway with optimal score. To solve this problem, [5] provides a randomized algorithm with high probability of finding the optimal pathway. Their algorithm applies to the scoring function that only considers adjacent proteins.

In this subsection, we extend their algorithm to handle our scoring function that also considers Level-2 neighbours. Let k be the length of the pathway. For any two proteins x_1, x_2, we assume that $s_1(x_1, x_2)$, which is the score for this protein pair if they are adjacent in the pathway, is given. Similarly, we assume that $s_2(x_1, x_2)$, which is the score for this protein pair if they are Level-2 neighbours, is also given. For a pathway $P = x_1, x_2, \ldots, x_k$, the score of P is $\sum_{i=1}^{k-1} s_1(x_i, x_{i+1}) + \sum_{i=1}^{k-2} s_2(x_i, x_{i+2})$.

We now present the randomized algorithm for computing the pathway. Let X be the set of input proteins. For every $x \in X$, we independently assign a color $c(x)$ drawn uniformly at random from the set $\{1, 2, \ldots, k\}$. A pathway is called *colorful* if it contains exactly one protein of each color. Let $I \subseteq X$ be the set of proteins that start the pathway. The algorithm finds the colorful path of maximum score that starts with a protein in I to every protein x.

For each nonempty set $T \subseteq \{1, 2, \ldots, k\}$ and for any two proteins x_1, x_2 such that $c(x_1) \neq c(x_2)$ and $c(x_1), c(x_2) \in T$, let $M(x_1, x_2, T)$ be the maximum score of a colorful pathway of length $|T|$ that starts with a protein in I, visits a protein of each color in T, with x_2 and x_1 as the second last and the last proteins, respectively. We can compute M recursively based on the following recurrence.

Recurrence (for $|T| > 2$):

$$M(x_1, x_2, T) = \max_{y: c(y) \in T - \{c(x_1)\} - \{c(x_2)\}} M(x_2, y, T - \{c(x_1)\}) + s_2(y, x_1) + s_1(x_2, x_1).$$

Base case (for $|T| = 2$):

$$M(x_1, x_2, \{c(x_1), c(x_2)\}) = \begin{cases} s_1(x_1, x_2) & \text{if } x_2 \in I \text{ and } c(x_1) \neq c(x_2) \\ \infty & \text{otherwise} \end{cases}$$

The time complexity of the algorithm is $O(2^k \times n^3)$ where n is the number of input proteins. Note that whether the algorithm can locate the pathway of optimal score depends on the step of assigning colors to the proteins. The probability of getting the optimal answer for each trial is $\frac{k!}{k^k}$. If we require this probability to be 99%, the number of trials should be at least $\frac{\ln(0.01)}{\ln(1-\frac{k!}{k^k})}$. For example, if $k = 7$, the number of trials should be greater than 750. Note also that the algorithm can be easily modified to report the top 100 pathways.

3.3. *Cleansing of gene expression data*

It is natural that the gene expression data may contain noise. Also, the gene expression data we obtained may not be directly related to the pathways we consider. We propose a simple greedy heuristics approach to remove some of these noise before the data is used for predicting the pathway. The principle behind our heuristics is as follows. For every pair of proteins, if a particular expression data item is noise, then by removing it, the absolute value of the correlation coefficient should not be decreased. So, we try to remove a certain percentage of expression data items so that the remaining data items can give the highest correlation coefficient of this pair.

We repeat the following procedure for every pair of proteins. To start with, let the number of data items be m. We discard one data item such that the remaining $m - 1$ data items can give the highest absolute value of the correlation coefficient. Then, we repeat the above step until a certain percentage of data items have been removed. In our experiments, we have about 350 data items for each protein and we have tried to discard 10% and 20% data items before we use it to reconstruct the pathway. In both cases, the results are similar and show a significant improvement in the accuracy of the predicted pathway.

In Table 2, we show the result of removing 10% gene expression data. The column "LZ Approach" shows the ranking of the true pathways by using the whole set of gene expression data. Comparing the results in this column with the one after removing 10% noise from the gene expression data, we found that the ranking of the true pathways in the latter case are all within 50 except one case that has a rank of 62. On the other hand, if we use the whole set of gene expression data, almost half of the true pathways are of rank more than 50. Five of them even have ranks of more than 100. We also have improvement in our approach after applying this cleansing procedure.

4. Experiments

We have conducted a number of experiments to verify the effectiveness of our approach. We make use of the 20 MAPK pathways of yeast in our experiments (see Table 1, the last column in the table shows the total number of permutations we have to consider if we use a brute-force approach to check each possible ordering). In all our experiments, we set tp $= 0.8$ and fn $= 0.1$ for our scoring function. In fact, we have tried other values such as tp $= 0.7$ and fn $= 0.2$. The results are similar.

Without Noise:
We first consider the restricted version of the problem, that is, the input proteins are all involved in the pathway. Table 2 shows the comparison result between LZ approach and our approach. Using our approach, about half of the true pathways appear among the top five answers and all the true pathways appear in the top ten answers except one that has a rank of 11. On the other hand, for LZ approach, only 7 out of 20 true pathways appear in the top ten answers. Some of the others even have ranks of more than 40. So, our approach is more accurate than the LZ approach.

With Noise:

For each pathway for testing, apart from the proteins that are involved in the pathway, we randomly select some other proteins of yeast to act as noise so that the input set is of size 30. Note that the amount of noise is about double the number of proteins in the pathway.

Since it may take a long time if we try all possible ordering, so we implement the randomized algorithm to use our scoring function as well as the scoring function used in the LZ approach for comparison. We have repeated the randomized algorithm for each pathway a number of times so that the probability for getting the optimal answer is about 99%. For examples, for a pathway of length 11, we have repeated the randomized algorithm for about 5000 trials. And we report the top 100 best pathways in each test case.

Table 3 shows the ranking of the true pathways using our approach and the LZ approach. Note that "n/a" refers to the case that the true pathway does not appear in the top 100 answers. From the experimental results, we can see that the LZ approach is very sensitive to noise. The true pathway does not appear in the top 100 answers almost for all cases. On the other hand, for our approach, we can see that among 20 cases, most of the cases have the true pathway in the top 15 positions. The results are slightly worse than the case without noise. Even for the cases that we cannot locate the true pathway within the top 100 answers, we found that there is a pathway with rank 1 or 2 that differs from the true pathway by only one protein. For the LZ approach, for some of the "n/a" cases, such a similar pathway can only be found with rank greater than 40. So, to conclude, it seems that using Level-2 neighbours is promising.

Table 3. Comparison Results with Noise in the Input Protein Set (Note that n/a means that the true pathway does not appear in the top 100 answers).

Index	LZ Approach	Our Approach
1	n/a	13
2	n/a	11
3	n/a	5
4	n/a	8
5	n/a	10
6	n/a	6
7	n/a	28
8	n/a	24
9	n/a	5
10	n/a	47
11	n/a	10
12	n/a	5
13	n/a	17
14	n/a	11
15	n/a	n/a
16	n/a	n/a
17	1	1
18	1	1
19	n/a	n/a
20	75	1

Note that the actual running time of the randomized algorithm depends on the length of the pathway. For path length of 11, it takes about 70 hours to complete all trials. Although it is faster than the brute-force approach for checking all possible pathways, it is desirable to derive a faster algorithm to solve the problem.

5. Conclusion

In this paper, we propose to make use of the information of Level-2 neighbours when computing the pathway for a set of proteins. Experiments show that our scoring function is more accurate. Also, we have extended the randomized algorithm in [5] to handle our scoring function.

Future directions include the followings. Besides, the gene expression data and the DIP, we should also consider using the information of protein domains and check if this can give a more accurate result. Also, pathways can have other forms such as a tree structure or a directed acyclic graph. Another direction is to derive better computational approaches to reconstruct these types of pathways. In fact, the authors in [5] also provide randomized algorithms to compute pathways that are in the form of rooted trees and two-terminal series-parallel graphs. From a computational point of view, whether we can improve the time for computing the optimal pathway and have a better approximation algorithm for solving the problem are also interesting directions to investigate.

References

1. Shawn M. Gomez, Shaw-Hwa Lo, and Andrey Rzhetsky. Probabilistic prediction of unknown metabolic and signal-transduction networks. *Genetics*, 159(3):1291–1298, 2001.
2. T.R. Hughes, M.J. Marton, A.R. Jones, C.J. Roberts, R. Stoughton, C.D. Armour, H.A. Bennett, E. Coffey, H. Dai, and Y.D. He. Functional discovery via a compendium of expression profiles. *Cell*, 102:109–126, 2000.
3. Yin Liu and Hongyu Zhao. A computational approach for ordering signal transduction pathway components from genomics and proteomics data. *BMC Bioinformatics*, 5:158, 2004.
4. C.J. Roberts, B. Nelson, M.J. Marton, R. Stoughton, M.R. Meyer, H.A. Bennett, Y.D. He, H. Dai, W.L. Walker, T.R. Hughes, M. Tyers, C. Boone, and S.H. Friend. Signaling and circuitry of multiple MAPK pathways revealed by a matrix of global gene expression profiles. *Science*, 287:873–880, 2000.
5. Jacob Scott, Trey Ideker, Richard M. Karp, and Roded Sharan. Efficient algorithms for detecting signaling pathways in protein interaction networks. In *Proceedings of the 9th Annual International Conference on Research in Computational Molecular Biology (RECOMB 2005)*, pages 1–13a, 2005.
6. Martin Steffen, Allegra Petti, John Aach, Patrik D'haeseleer, and George Church. Automated modelling of signal transduction networks. *BMC Bioinformatics*, 3:34, 2002.
7. I. Xenarios, L. Salwinski, X.J. Duan, P. Higney, S.M. Kim, and D. Eisenberg. DIP, the database of interacting proteins: a research tool for studying cellular networks of protein interactions. http://dip.doe-mbi.ucla.edu.

INVESTIGATING ROLES OF DNA FLEXIBILITY IN PROMOTER RECOGNITION AND REGULATION*

J. D. BASHFORD

School of Mathematics and Physics, University of Tasmania
Private Bag 37,
Hobart, Tasmania 7001, Australia
E-mail: James.Bashford@utas.edu.au

We attempt to partly quantify the "indirect readout" feature of promoter recognition which may enhance or inhibit binding of RNA polymerase. Using recent advances in understanding of B-DNA flexibility, we develop and test a toy model of DNA bending. The model is used to estimate the energetic cost of a specific kind of bend, simulating a DNA-protein binding event, at each site within a promoter-containing DNA sequence. The flexibility of strong, unregulated promoters from lytic coliphages is compared with that of weaker, unregulated *Escherichia coli* sequences. We find i) in both cases the favoured location for introduction of an RNAP-sized bend is often within the core promoter region although, proportionally, far more phage sequences have this feature; ii) promoters regulated by Crp (a DNA-bending transcription factor) lack this feature, being most flexible 50 to 80 sites upstream of initiation. These observations may suggest i) core flexibility in unregulated promoters enhances transcription activity and ii) differing roles for flexibility in Crp-regulated versus unregulated sequences in activation.

1. Introduction

Helical curvature and flexibility are key to many biological functions of DNA, including transcription, replication and protein-binding in addition to packaging within the cell. A substantial body of evidence, accumulated from experiment[1–6] and molecular dynamical simulation[7–9] is consistent with the idea that B-DNA flexibility[7] "...can be understood as a combination of small geometrical distortions at the constituting steps." For a recent review see Ref. 10. An intriguing corollary[11] of this observation is the potential existence of structural and mechanical "codes" underlying regulation of gene expression, which might be profitably used for analysis of existing repositories of genomic data.

Theoretical investigations including consideration of sequence-dependent DNA flexibility to date include: an "atlas" of DNA mechanical properties within the *Escherichia coli* chromosome[12], DNA looping[13] and supercoiling-dependent transcription regulation[14]. In a similar vein, prokaryotic and eukaryotic promoter sequences have also been shown[15–17] to have distinct sequence-dependent structural/mechanical properties which may assist in promoter location by RNA polymerase (RNAP)[18].

*To appear in proceedings of 3rd RECOMB Regulatory Genomics Workshop, Singapore 17–18 July, 2006.

One potential shortcoming of investigating large sets of promoter sequences[16,17] which the present paper attempts to redress, is their generality: the averaged properties of a large set of promoter sequences need not be representative of many individual sequences. With the advent of encyclopaedic databases such as EcoCyc[19] (http://ecocyc.org) it has become possible to implement more restrictive classification schemes for sets of promoter sequences in order to address such questions.

Since transcription regulation is also frequently governed by DNA bending and qualitatively similar deformations are likely to occur in isolated B-DNA and DNA-protein complexes[7], it is natural to question whether variations in flexibility of promoter sequences might also correlate with the efficiencies of RNAP or transcription factor (TF) binding to DNA. In particular we investigate variations on the scale of DNA-protein complexes, in order to test the hypothesis that the affinity of a binding site correlates with the susceptibility of surrounding sequence to deformation. That is, we seek to quantify, partially at least, "indirect readout" which may enhance or inhibit the recognition of binding sites by their cognate proteins.

The structure of the paper is as follows: First a toy model of DNA bending is introduced and its robustness is tested upon (unregulated) promoters found in the T7 bacteriophage genome. The resulting observations lead us to develop a method of analysing larger sets of sequences, which is subsequently applied to sets of both regulated and unregulated bacterial promoters in order to illustrate differences in mechanical properties.

2. Materials and Methods

2.1. *Promoter sequences*

The genomes for *E. coli* strain K12 plus seven members of the T7 phage superfamily were downloaded from Genbank (http://www.ncbi.nlm.nih.gov) and, for unannotated sequences, promoter locations obtained from the literature as indicated in Table 1. While there are nu-

Table 1. Genomes used in the study.

Genome	Accession No.	Reference	Genome	Accession No.	Reference
T7	**NC001604**		T3	**NC003298**	
A1122	**NC004777**	20	YeO3-12	**NC001271**	
gh-1	**NC004665**		SP6	**NC004831**	21
K1-5	**AY370674**	21	*E. coli*	**NC000913**	19

merous mechanisms of transcription regulation (see Ref. 22 for a recent review) within a DNA-bending model we are limited to consideration of those related to structural deformation of DNA, for example the "indirect readout" of promoter sequences (or possibly binding sites for upstream-binding TFs). Hence for the purposes of this study we shall refer to a promoter as "unregulated" simply if no TF binding sites are present.

2.2. *DNA flexibility model*

The model of DNA we use, of a weakly-bending rod is qualitatively similar to ones used in other studies of dynamical, DNA duplex flexibility (e.g. Refs. 1, 4) with one key difference. Usually such flexibility is investigated with respect to thermal fluctuations, where a full, statistical-mechanical treatment is appropriate. By contrast, we wish to consider only conformational changes in DNA likely to be similar to those induced by protein binding. In the presence of a protein, certain DNA conformations would be expected to become strongly preferred, allowing simplification of any full, statistical model. It is this rationale which forms the basis of our investigation below.

In the weakly-bending rod picture one assumes the DNA duplex has first-order elasticity and bending is isotropic. Then the bending energy of a DNA segment containing $N = 2s + 1$ internal steps is given by

$$\mathcal{E} = \frac{1}{2} \sum_{j=-s}^{s} k_j \chi_j^2 , \qquad (1)$$

where k_j and χ_j are the (apparent) stiffness and curvature associated with the jth internal step. The generalisation to segments of even length is obvious. Following Ref. 1 we may factorise the stiffness

$$k_j = \frac{RTP}{\ell} f_j^{-1} , \qquad (2)$$

where f_j are flexibility parameters associated with each step, P, is the persistence length of B-DNA and ℓ is the size of each step. Let us now assume, for simplicity, the bend induced

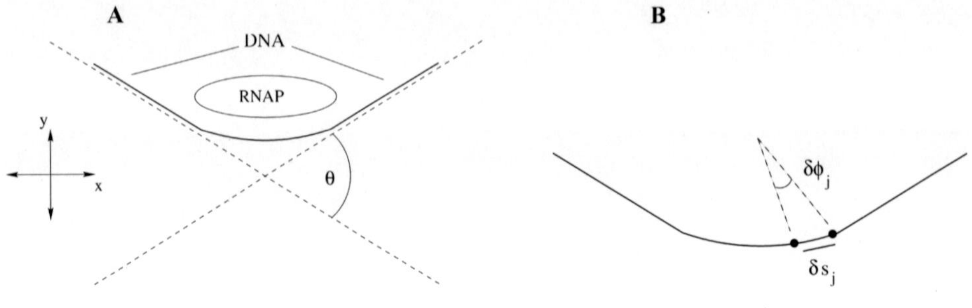

Figure 1. **(A)** Hyperbolic deformation of B-DNA axis by RNAP. **(B)** Curvature of jth hyperbolic segment.

in DNA has an analytic expression. In particular we consider a hyperbolic curve, sketched in Figure 1A, interpolating between two linear regions, which depends on two parameters: θ, the angle through which the B-DNA helix axis is bent and L, a quantity determining the size of the bent region, which will be related to the protein-binding "footprint" size below. Consider, initially,

$$F(z) = \left(\sinh(z/L), \tan\frac{\theta}{2} \cosh(z/L) \right) ,$$

for some parameter z with a *fixed* range. Formally the curvature, χ_j associated with a segment of F is written in terms of tangential angle subtended, $\delta\phi_j$, and segment length, δs_j, as in Figure 1B

$$\begin{aligned}\chi_j &= \frac{\delta\phi_j}{\delta s_j}, \\ &\equiv \frac{\delta\phi_j/\delta j}{\delta s_j/\delta j},\end{aligned}$$

where δj defines the size of an internal step. In the continuum limit, ($\delta j \to 0$), for the hyperbola (3) one obtains the expressions

$$\frac{d\phi}{dt} = \tan\frac{\theta}{2}\left(\cosh^2 t + \tan^2\frac{\theta}{2}\sinh^2 t\right)^{-1}, \quad (3)$$

$$\frac{ds}{dt} = \sqrt{\cosh^2 t + \tan^2\frac{\theta}{2}\sinh^2 t}\,. \quad (4)$$

For a model DNA strand of N basepairs, the second of these expressions can then be used to partition F into $(N-1)$ segments of equal length corresponding to internal dinucleotide steps. The integral in Eq. (4) is readily evaluated as

$$I(z) \equiv \int_0^{z/L} dt\frac{ds}{dt} = -iE(iz/L, \sec\theta/2)\,, \quad (5)$$

where E is the elliptic integral of the second kind[23] and $i^2 = -1$. In the case where the bend is localised around a central dinucleotide step and N is even, there are $N/2-1$ steps upon either side of the origin. If we have a fixed range ($-b \leq z \leq b$) for some real, positive b, then the total segment length is

$$L_T = 2I(b)\,,$$

and hence the step size should be

$$\nu = L_T/(N-1)\,. \quad (6)$$

It follows that the numbers τ_j, satisfying

$$I(\tau_j) = j\nu;\ j = -(N/2-1), \ldots, N/2-1\,, \quad (7)$$

define hyperbolic segments of length ν. For example, $F(\tau_1)$ corresponds to the center of the first dinucleotide step downstream of the bend centre.

When the bend is centred upon a single base pair and, for convenience we take N odd, there are $(N-1)/2$ steps upon either side of the origin. With appropriately defined L_T and ν the corresponding τ_j are obtained from

$$I(\tau_j) = (j - \frac{1}{2})\nu;\ j = -(N-3)/2, \ldots (N-1)/2\,.$$

Substituting Eqs. (2)–(4) and (7) into the energy expression (1) we obtain the main formula of this paper. The energy required to create a bend with dimensions (L, θ) in the center of a given DNA sequence is:

$$\frac{1}{\kappa}\mathcal{E}(L,\theta) = \sum_j f_j^{-1}\chi_j^2\,, \tag{8}$$

$$\kappa = \frac{RTP}{2\ell}\,,$$

$$\chi_j = \tan\frac{\theta}{2}\left(\tan^2(\frac{\theta}{2})\sinh^2\tau_j + \cosh^2\tau_j\right)^{-3/2}\,, \tag{9}$$

where the summation limits on j depend on whether the bend is localised around a single base pair or dinucleotide. Implicit in this generalisation is the assumption that we are free to choose whether N is odd or even. In the context of the *E. coli* genome analysis below, N determines a neighbourhood of a far larger DNA strand and, provided N is much larger than the central, bent region determined by L, this assertion is justified. The basic entity used in the remainder of this paper is an energy "landscape", generated by plotting the right-hand side of Eq. (8) versus position, X within a given genomic region upon which, e.g., sites of transcriptional significance can be superimposed.

It is clear that to simulate a protein-binding event with characteristic footprint size Λ we need a further constraint upon the bend size parameter L. This can be accomplished by considering the full width at half-maximum (FWHM) value X_{FWHM} of the bell-shaped curve (9)

$$X_{FWHM} = \text{ArcSinh}\frac{\sqrt{(2\tan\theta/2)^{2/3} - 1}}{\sec\theta/2}\,.$$

Then Λ is simply the size of the central subset of values τ_j, that is:

$$\Lambda = |\{\tau_j : \tau_j \leq X_{FWHM}\}|\,. \tag{10}$$

2.3. *Flexibility parameters*

We investigate the behaviour of the model with respect to four sets of flexibility values, f_j^{-1}, derived from different experimental approaches.[1,3,4,24] Sets GT and FK are derived from nearest-neigbour dimer stacking energies. However the former is obtained from DNA melting experiments and therefore includes H-bond denaturation in addition to stacking. In contrast, parameter set FK is taken from a study of the electrophoretic mobility of nicked B-DNA[3] and hence can more reasonably be assumed to consist purely of stacking effects. For both sets the flexibility parameters were obtained via

$$f_j^{-1} = \Delta G_j / \sum_{j=1}^{10} \Delta G_j\,,$$

where ΔG_j is the stacking free energy of dimer j. The set MD is obtained from a molecular dynamics simulation[24] of base-pair step deformation. The isotropic "stiffness", κ_j, of dimer j is approximated by the harmonic mean of the force constants associated with roll-roll and tilt-tilt deformations (Table 1 of Ref. 24). The values for MD are then obtained from

$$f_j^{-1} = \kappa_j / \sum_{j=1}^{10} \kappa_j \, .$$

Table 2. Dinucleotide step flexibility parameters f_j^{-1}.

Model	AA	AT	TA	GG	GC	CG	AG	AC	GA	CA
GT[1]	0.703	0.854	0.615	0.984	1.792	1.124	0.78	1.323	1.230	0.790
FK[3]	0.924	1.116	0.158	1.199	1.806	0.758	0.883	1.507	1.191	0.458
MD[24]	1.062	1.182	0.636	1.109	1.182	0.716	1.025	1.178	0.996	0.800
OK[4]	1.163	1.111	0.714	0.990	-*	0.541	0.769	2.222	1.449	0.602

Lastly, we can tentatively include set OK[4], derived from an empirical study of DNA bending dynamics at the submicrosecond scale. This last set is problematic because the inverse stiffness (the analogue of f_j for MD) is the fundamental quantitity extracted from empirical data. Unfortunately the value $f_{GC} = 0$, making realistic comparisons between predictions of this dataset and the other three unfeasible. *One possibility is to acknowledge the stiffness for dimer GC is much greater than other dimers and arbitrarily assign it a value such that landscapes generated are of the same order of magnitude as those from other datasets. To this end we vary $0.1 \leq f_{GC}^{-1} \leq 0.4$, however it is clear lesser significance should be attached to results thus obtained.

3. Results

Protein-DNA interactions are governed by Boltzmann kinetics and formally, a statistical mechanical treatment is required. It is simpler, however, to assume the bend parameters θ, L correspond to average values, associated with a highly-favoured conformation, and consider fluctuations about this. The effects of varying these parameters will be considered in a subsequent section.

3.1. *T7 Promoter sequences are associated with high flexibility*

For our initial investigation of prokaryotic promoters we select two sequences from the T7 coliphage genome: T7A1, one of the strongest known and most widely studied *E. coli* RNAP-binding promoters; ϕ1.1A, one of a set of sixteen, highly similar[29] pharge RNAP-specific promoters in the T7 genome. We choose promoters binding to nonhomologous RNAP species in order to quantify the generality of any regularities seen in bending energy profiles.

In the standard model of *E. coli* promoter activation there are three main steps: rapid, initial binding of RNAP to promoter DNA; slow isomerisation of the closed complex and DNA melting. Hence there are several conformational changes which bending deformations could be associated with. In addition to -35 and -10 hexamers which closely match their consensus motifs[25], the A1 promoter contains a large UP element[26] extending from sites -42 to -80. Sclavi *et al.*[27] suggested that at least three intermediate structures exist during the initial phase of activation at the T7 A1 promoter, with the template strand protected from regions -55 to -43, -45 to -31 and -55 to -21 respectively, however corresponding regions on the non-template strand were unable to be assigned. In the present model, since DNA strands are not distinguished the above delineations should be considered as lower bounds on L. In modelling RNAP-DNA complexes we therefore consider values of Λ ranging from 13 (initial binding lower bound, above) to 75 (a full closed complex, from -55 to +20) base pairs. Further we can distinguish an "upstream" region ranging from sites -80 to -36, associated with initial RNAP-DNA contacts, from the "core" promoter (sites -35 to +20) which is contacted by RNAP during later steps in the activation pathway. Similarly we consider θ values in the range $10°$ (assumed) to $60°$ for a closed complex[a].

The activation pathway of T7 RNAP is less well known, however the closed complex is associated[28] with a"footprint" extending from sites -17 to +6 overlapping with a base sequence which is highly conserved in the genomes of T7 and other members of the *Podovirales* T7 superfamily.[29] Further the axial bend is estimated from crystal structures[30] to be $\theta \simeq 80°$.

Figure 1 shows normalised $(\mathcal{E}/\Lambda)$ A1 and $\phi 1.1A$ landscapes generated by Eq. (8) using the mid-range values of Λ, θ and each set of flexibility parameters. In each case the origin represents the transcription initiation site. The variations due to different choices of f_j^{-1} are qualitatively similar for the empirically-derived sets FK, GT and OK: extrema of the dotted and solid curves are generally in one-to-one correspondence, differing largely in amplitude, while the *ab initio* MD parameters yield far less variation. From Fig 1A it is seen that curves FK, GT and OK predict global bending energy minima coinciding with the A1 UP element while MD places it in the core region, distal to the -10 element. Further a second major bacterial promoter, A2, having initiation site at +128 is characterised, in all 4 landscapes, by a broad local minimum in the core region.

3.2. *Effects of varying L, θ*

The effect of varying bend parameters L and θ can be summarised upon inspection of Eq. (9), which describes a bell-shaped function. In fact Eq. (8) can be likened to a statistical moment of flexibility with the values of the curvature, χ, providing the weighting factors at each step. As θ increases towards the pathological limit of $180°$ or L decreases, the width of the bell curve tends to zero, leading to an increasingly "jagged" landscape. Conversely for small θ or large L the landscape becomes smooth and homogeneous. This effect is illustrated in Figure 3, where A1 landscapes are plotted for footprints ranging in size from 10-40 bp. The common features include regions of flexibility, localised in the central UP region of A1 and core of A2, plus relative rigidity just downstream of initiation. In the

[a]There is strong evidence[31] that *E. coli* RNAP "wraps" DNA around itself, through an angle of $\sim 300°$, rather than bending the helix axis through $\sim 60°$. Equation (8) is invariant under swapping $\theta \leftrightarrow 360 - \theta$, Thus our method does not distinguish the two scenarios.

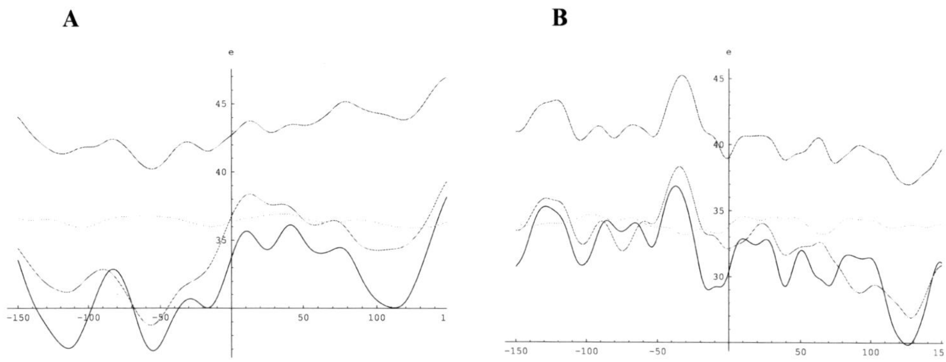

Figure 2. Energy landscapes for **(A)** T7 A1 ($\Lambda = 35$, $\theta = 30°$) Black, red, green and blue curves refer to flexibility parameter datasets FK, GT, MD and OK respectively.

"small-to-moderate" bend regime defined by $10° \leq \theta \leq 120°$ we find the main effect of changing θ is a small overall scaling, with little discernible sensitivity (data not shown) of landscape shape. Generally speaking the robustness of a peak/trough is a function of its amplitude and width. Bearing in mind a Boltzmann-kinetic description of DNA-protein

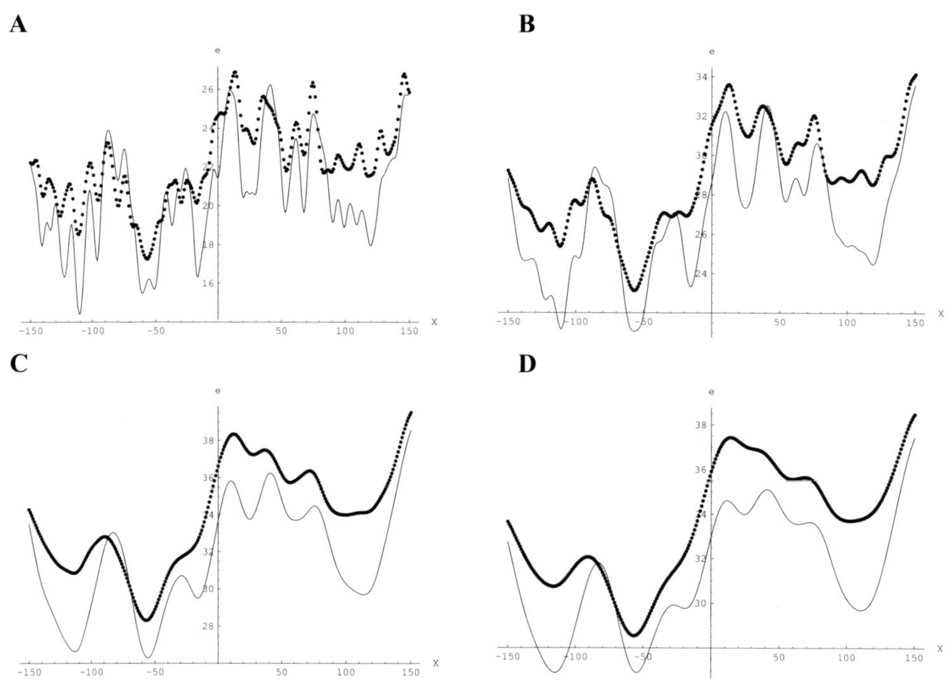

Figure 3. **(A)-(D)** Energy landscapes for T7 promoter A sequence corresponding to $\Lambda = 10, 20, 30, 40$ bp and $\theta = 60°$. Solid and dotted curves refer to flexibility parameter datasets FK and GT respectively.

binding we conjecture that the robustness of a given binding site extremum with regard to fluctuations in L and θ may correlate with binding rates. Such a treatment is the subject of ongoing investigation.

3.3. *Promoter comparison*

Having developed a basic model of promoter flexibility it is important to consider comparisons of different promoter sequences. For the purposes of this preliminary study we shall pursue a very simple strategy, to illustrate the kinds of insights which may be obtained via this approach. Specifically we shall look at the position of the global bending energy minimum within a promoter-containing region of DNA.

Based upon our observations of host- and phage-specific promoters in the T7 genome (data not shown) we will focus on the former, on the grounds that the global (within the region) minimum appears to correlate better with the core sequence for *E. coli* RNAP (5 of 7 sequences), than its phage counterpart (2 of 16 sequences). Further we shall use the FK set of flexibility parameters in preference over GT, MD and OK using the following rationale: All three empirically-derived data sets were demonstrated above to give qualitatively similar landscape shapes, hence the theoretical set, MD, might be rejected. Furthermore FK could be argued superior to GT on the grounds that it reflects "pure" stacking contributions, although there is some controversy over whether stacking correlates at all with dimer flexibility.[10] As mentioned in the Materials and Methods, the value of f_{GC}^{-1} in set OK was modified from the empirically-obtained zero value, thus lesser significance should be attributed to resulting predictions, particularly in GC rich areas. All landscapes below were calculated for 300bp regions, using the mid-range values of $L = 35$ bp and $\theta/2 = 30°$, used previously for the T7 genome.

The T7 genome contains three strong (A1, A2, A3) plus four weak (D,B,C and E) *E. coli* promoters, the latter set having no known *in vivo* function. Other members of the T7 superfamily contain varying numbers of similar major and minor promoters. The energy and location of the "global" minimum of each promoter sequence in the seven T7-like phages of Table 1 is plotted in Fig. 4A, where we have distinguished strong promoters (dots) from weaker ones (circles). It is seen that there is little correlation between bending energy and promoter activity. However in Fig. 4B one sees a tendency for minima to cluster around locations of significance. Many sequences have maximum flexibility within the so-called "spacer" region, between the -35 and -10 hexamers, while a second cluster occurs between -48 to -66, in the region of the UP element (where present). Other clusters are at least partially an artefact of window size: in several instances the window centred upon one promoter contains a deeper minimum up- or down-stream, associated with another promoter. Our initial observations thus suggest correlation between global bending energy minima and UP- (where present) or core-binding regions of promoter sequences. On the other hand the T7-like phages possess some of the strongest known *E. coli* promoters and, having lytic life cycles, constitute rather pathological examples of regulatory networks. It is clearly important to analyse promoters from other genomes in the same way.

To this end we next studied two sets consisting of i) 125 unregulated and ii) 200 regu-

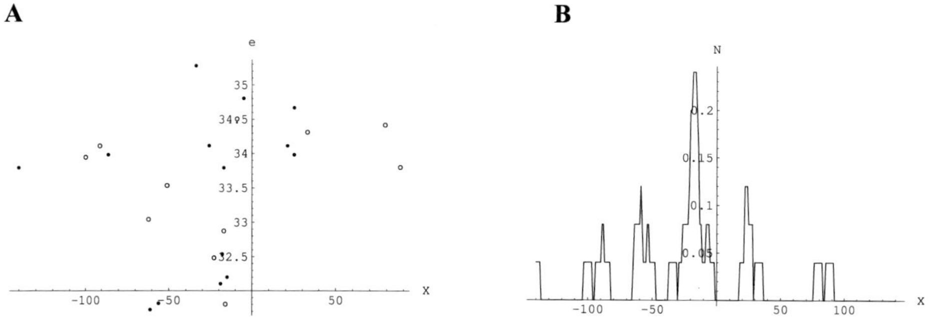

Figure 4. **(A)** Energy of sequence global minimum plotted against location for T7 host-specific promoters. Dots and circles refer to major and minor promoters respectively. **(B)** Proportion, N, of T7 sequences having a global minimum at within a hexamer region centred on location X.

lated promoters from the *E. coli* genome. In the latter set we did not distinguish between number or type of TF binding sites. Unregulated promoters (dotted curve; Fig. 5A) have a weak preference for maximal flexibility proximal to the initiation site (-3 to -10) or distal, around -60 to -70. These peaks are similar to those in Fig. 4B for the T7 family although the extent of clustering is significantly less. The regulated promoters (solid curve, Fig. 5A) are also characterised by a peak in the proximity of the initiation site. It follows that some regularities are apparent for large promoter sets, such as the union of the two plotted in Fig. 4A, consistent with findings in other studies.[16,17] However this raises the question

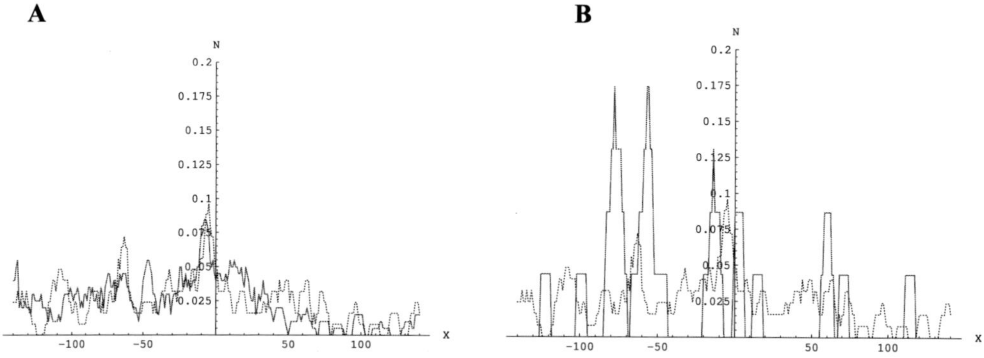

Figure 5. **(A)** Comparison of distributions of global minimal flexibility for families of unregulated (dotted) and randomly chosen regulated (solid) *E. coli* promoters. **(B)** Similar comparison for unregulated (dotted) and Crp-only regulated (solid) promoters.

as to whether the indiscriminate grouping of all kinds of regulated promoters "averages" away more underlying trends associated with more specific classifications of promoters. Indeed, in a study of DNA flexibility we should restrict our analysis to TFs which bend the B-DNA helix. The global regulator Crp is a natural candidate as it is known to induce bends in DNA and the EcoCyc database lists 23 promoters which are controlled solely by Crp, provided we do not distinguish promoters with single or multiple sites. The clustering plot for Crp-only promoters contrasts markedly with that of the set of unregulated *E. coli* promoters in Figure 5B, which shows the former have a preference for global minima in either the -50 to -60 or -70 to -80 regions, although a smaller peak occurs proximal to the -10 element No correlation of minimal flexibility with Crp binding sites (data not shown) was observed however.

4. Discussion

In this study we have developed a simple model of DNA bending and applied it to sequence-dependent variations in flexibility of promoter containing regions of DNA. Quite generally for *E. coli* promoters in lytic phages, the most energetically-favoured site for introduction of an RNAP-sized bend was often found to lie within the core promoter, in the spacer region from -20 to -10 (Figure 4B). Similar, albeit greatly reduced trends were seen for unregulated host promoters which, in turn contrasted greatly with the characteristic flexibilty profile found for for Crp-regulated promoters (Fig 5B). It is tempting to speculate that, in the absence of activator transcription factors, flexibility in these core promoter regions may be beneficial to the later stages of activation[27]: possibly recognition of the -10 element or isomerisation from open to closed transcription complex.

Predictions for local bending energy based upon three sets of empirically-derived dimer flexibility parameters[1,3,4] were found to be inqualitative agreement with each other but less so with a recent *ab initio* model.[24] Variation of the bend parameters θ, L demonstrated robustness of landscape extrema which were broad and deep, such as the minima associated with core promoter regions, necessary in a full statistical mechanical model.

There are obvious criticisms of the both the bending model and subsequent sequence analysis, which we now discuss. Firstly, we assumed DNA bending was isotropic, smooth and planar in nature, all of which are highly idealised conditions. For example, while the bending of the helix axis in a Crp-DNA complex is approximately planar[9] it is also highly localised within two "kinked" dimers. Moreover due to the anisotropy between roll and tilt degrees of freedom, continuous bends commonly writhe: planar bends of this nature tend to be rare in DNA-protein complexes. This is one possible reason why correlations between other binding sites (e.g. for Crp) and landscape extrema may not be seen. On the other hand flexibility in the α-CTD binding region could enhance activation complex formation by facilitating Crp-RNAP[32] rather than direct Crp-DNA binding.

Finally the stacking-derived dinucleotide step parameters FK and GT are unable to distinguish intrinsic and dynamical curvature, both of which play different roles in promoter activation.[33] One possible improvement to rectify this may be to use empirically-derived steps of bending propensity such as the trinucleotide set of Ref. 5.

The shortcomings of the crude "global" minimum strategy adopted in this study are readily apparent from the landscape in Figure 1D: the equal-deepest minimum is located upstream of the core promoter sequence in a region which contacts the α- subdomain of RNAP. However there is an equally deep minimum appears at -115, removed from any known sites of transcriptional significance. It is clear that more sophisticated consideration of landscape features, such as region size, maxima near initiation or "competition" between neighbouring minima, are required to fully investigate comparisons of promoter strength.

While the model of DNA-RNAP binding developed here is clearly unsatisfactory in many ways, the fact that we observe mechanical regularities among different classes of promoter sequences suggests that some underlying aspect of promoter recognition has been captured. Moreover this aspect is mechanical and complementary to the usual approach of searching for conserved base motifs. It is our hope that, ultimately quantitative, as well as qualitative, information regarding DNA-protein binding sites might be extracted from genomic data. This work in progress is intended to be a tentative step in investigating the feasibility of this proposition.

Acknowledgments

This research was funded by Australian Research Council grant DP0344996. The author wishes to thank I. Molineux for helpful discussions during this work and an anonymous referee for suggesting improvements.

References

1. Scipioni, A., Anselmi, C., Zuccheri, G., Samori, B. and De Santis, P. (2002), Sequence-dependent DNA curvature and flexibility from scanning force microscopy images, *Biophys. J.* **83**, 2408–2418.
2. Gotoh, O. and Tagashira, Y. (1981), Stabilities of nearest-neighbour doublets in double-helical DNA determined by fitting calculated melting profiles to observed profiles, *Biopolymers* **20**, 1033–1042.
3. Protozanova, E., Yakovchuk, P. and Frank-Kamenetstkii, M. D. (2004), Stacked-unstacked equilibrium at the nick site of DNA, *J. Mol. Biol.* **342**, 775–785.
4. Okonogi, T. M., Alley, S. C., Reese, A. W., Hopkins, P. B. and Robinson, B. H. (2002), Sequence-dependence dynamics of duplex DNA: the applicability of a dinucleotide model, *Biophys. J.* **83**, 3446–3459.
5. Brukner, I., Sánchez, R., Suck, D. and Pongor, S. (1995), Sequence-dependent bending propensity of DNA as revealed by DNase I: parameters for dinucleotides, *EMBO J.* **14**, 1812–1818.
6. Kanhere, A. and Bansal. M. (2003), An assessment of three dinucleotide parameters to predict DNA curvature by quantitative comparison with experimental data, *Nucl. Acids Res.* **31**, 2647–2658.
7. Pèrez, A., Noy, A., Lankas, F., Luque. F. J. and Orzoco, M. (2004), The relative flexibility of B-DNA and A-RNA duplexes: database analysis, *Nucl. Acids Res.* **32**, 6144–6151.
8. Olson, W. K., Gorin, A. A., Lu, X.-J., Hock, L. M. and Zhurkin, V. B. (1998), DNA sequence-dependent deformability deduced from protein-DNA crystal complexes, *Proc. Natl. Acad. Sci. USA* **95**, 11163–11168.
9. Dickerson, R. E. (1998), DNA bending: the prevalence of kinkiness and the virtues of normality, *Nucl. Acids. Res.* **26**, 1906–1926.
10. Lankaš, F. (2004), DNA sequence-dependent deformability–insights from computer simulations, *Biopolymers* **73**, 327–339.
11. Thayer, K. M. and Beveridge, D. L. (2002), Hidden Markov models from molecular dynamic simulations on DNA, *Proc. Natl. Acad. Sci. USA* **99**, 8642–8647.
12. Gorm Pedersen, A., Jensen, L., Brunak, S., Staerfeldt, H. and Ussery, D. W. (2000), A DNA structural atlas for *Escherichia coli. J. Mol. Biol.* **299**, 907–930.
13. Becker, N. A., Kahn, J. D. and Maher, L. J. III (2005), Bacterial repression loops require enhanced DNA flexibility, *J. Mol. Biol.* **349** 716–730.

14. Hatfield, G. W. and Benham, C. J. (2002), DNA topology-mediated control of global gene expression in *Escherichia coli. Annu. Rev. Genet.* **32**, 175–203.
15. Wang, H., Noordewier, M. and Benham, C. J. (2004), Stress-Induced DNA duplex destabilization (SIDD) in the *E. coli* genome: SIDD sites are closely associated with promoters, *Genome Research* **14**, 1575–1584.
16. Kanhere, A. and Bansal, M. (2005), Structural properties of promoters: similarities and differences between prokaryotes and eukaryotes, *Nucl. Acids Res.* **33**, 3165–3175.
17. Fukue, Y., Sumida, N., Nishikawa, J. and Ohyama, T. (2004), Core promoter elements of eukaryotic genes have a highly distinctive mechanical property, *Nucl. Acids Res.* **32**, 5834–5840.
18. Bashford, J. D. (2006), Salerno's model of DNA re-analyzed: could breather solitons have biological significance? *J. Biol. Phys.* **32**, 1–21.
19. Keseler, I. M., Collado-Vides, J., Gama-Castro, S, Ingraham, J., Paley, S., Paulsen, I. T., Peralta-Gil, M and Karp, P. D. (2005), EcoCyc: a comprehensive database resource for *Escherichia coli. Nucl. Acids. Res.* **33**, D334–337.
20. Garcia, E., Elliott, J. M. Ramanculov, E., Chain, P. S. G., Chu, M. C., Molineux, I. J. (2003), The genome sequence of *Yersinia pestis* bacteriophage ϕA1122 reveals an intimate history with the coliphage T3 and T7 genomes, *J Bacteriol.* **185**, 5248–5262.
21. Scholl, D., Kieleczawa, J. Kemp, P., Rush, J., Richardson, C. C., Mer, C., Adhya, S. and Molineux, I. J. (2004), Genomic analysis of bacteriophages SP6 and K1-5, an estranged subgroup of the T7 supergroup, *J. Mol. Biol.* **335**, 1–1171.
22. Browning, D. F. and Busby, S. J. W. (2004), The regulation of bacterial transcription initiation, *Nature. Rev. Microbiol.* **2**, 1–9.
23. Gradsteyn, I. S. and Rhyzhik, I. M., *Tables of Integrals, Series and Products*, 6th ed. Academic Press, New York, 2000.
24. Lankas, F., Sponer, J., Langowski, J. and Cheetham, T. E. III (2003), DNA basepair step deformability inferred from molecular simulations, *Biophys. J.* **85**, 2872–2883.
25. Lisser, D. and Margalit, H. (1993), Compilation of *E. coli* mRNA sequences, *Nucl. Acids Res.* 1507–1516.
26. Rao, L., Ross, W., Appleman, J. A., Gaal, T., Leirmo, S., Schlax, P. J., Record, M. T., Jnr and Gourse, R. L. (1994), Factor-independent actvivation of *rrnB* P1; an "extended" promoter with an upstream element that dramatically increases promoter strength, *J. Mol. Biol.* **235**, 1421–1435.
27. Sclavi, B., Zaychikov, E., Rogozina, A., Walther, F., Buckle, M. and Heumann, H. (2005), Real-time characterization of intermediates in the pathway to open complex formation by *Escherichia coli* RNA polymerase at the T7A1 promoter, *Proc. Natl. Acad. Sci. USA* **102**, 4706–4711.
28. Imburgio, D., Rong, K. Ma. and McAllister, W. T. (2000), Studies of promoter recognition and start site selection by T7 RNA polymerase using a comprehensive collection of promoter variants, *Biochemistry* **39**, 10419–10430.
29. Chen, Z. and Schneider, T. D. (2005), Information theory based T7-like promoter models: classification of bacteriophages and differential evolution of promoters and their polymerases, *Nucl. Acids Res.* **33**, 6172–6187.
30. Yin, Y. W. and Steitz, T. A. (2002), Structural basis for the transition from initiation to elongation transcription in T7 RNA polymerase, *Science* **298**, 1387–1395.
31. Rivetti, C. R., Guthold, M. and Bustamente, C. (1999), Wrapping of DNA around the *E. coli* RNA polymerase open promoter complex, *EMBO J.* **18**, 4464–4475.
32. Benoff, B. *et al.* (2002), Structural basis of transcription activation: the CAP-αCTD complex, *Science* **297**, 1562–1566.
33. Nickerson, C. A. and Achberger, E. C. (1995), Role of curved DNA in binding of *Escherichia coli* RNA polymerase to promoters, *J. Bacteriol.* **177**, 5756–5761.

REGULATORY NETWORKS OF GENES AFFECTED BY MORA, A GLOBAL REGULATOR CONTAINING GGDEF AND EAL DOMAINS IN *PSEUDOMONAS AERUGINOSA*[*]

WENG-KEONG CHOY[†]
School of Life Sciences & Chemical Technology
Ngee Ann Polytechnic
535, Clementi Road,
Singapore 599489

VLADIMIR B. BAJIC
South African National Bioinformatics Institute,
Private Bag X17
University of Western Cape, SANBI
Bellville 7535
South Africa

MOK-WEI HENG
Department of Biological Sciences
National University of Singapore
10 Science Drive 4
Singapore 117543

MERLIN VERONIKA
Knowledge Extraction Lab,
Institute for Infocomm Research,
Singapore 119613

SANJAY SWARUP
Department of Biological Sciences
National University of Singapore
10 Science Drive 4, Singapore 117543
Tel: (65) 6874 7933 Fax: (65) 6779 2486
EMAIL: DBSSS@NUS.EDU.SG

Pseudomonas aeruginosa is a well-known opportunistic bacterial pathogen, which causes high mortality in immuno-compromised patients. It is the main cause for increased morbidity and mortality in cystic fibrosis (CF) patients (Stover *et al.*, 2000) and in patients with AIDS-associated infections (Quinn, 1998). CF patients commonly suffer from respiratory tract infections by *P. aeruginosa*, which can lead to persistent infections of the lungs. Such patients eventually succumb to the lung damage inflicted by the persistent infection, resulting in pulmonary failure and death (Stover *et al.*, 2000). Reports of *P. aeruginosa* infections from hospital are widespread. A survey in the United Kingdom showed that the most common organism isolated from the sputum of CF patients is *P. aeruginosa* (Pitt *et al.*, 2003). In an Italian study, *P. aeruginosa* isolation

[*] This work is supported by Academic Research Fund, National University of Singapore.
[†] Weng-Keong Choy was funded by a research scholarship awarded by the National University of Singapore.

accounted for 14% of the 25,266 consecutive aerobic bacteria isolates and for 21% of the clinically important Gram negative bacteria isolates from patients (Bonfiglio *et al.*, 1998). Two main reasons have emerged for the persistence of *P.* aeruginosa in hospital infections and in CF patients leading to chronic situations; (i) the establishment of biofilms (surface-attached bacterial communities) in lungs of CF patients and (ii) emergence of multidrug resistant (MDR) strains of *P. aeruginosa*. Bacterial biofilms can cause serious environmental problems and cells may become 100- to 1,000-fold more resistant to antimicrobial agents. For these reasons, there is a great deal of interest worldwide in the development of antipseudomonal antibiotics that will be especially effective in combating biofilm formation. A novel regulator, *morA*, was previously identified to affect biofilm formation in *P. aeruginosa* (Choy *et al.*, 2004). Gene expression studies were subsequently done on the Wild-Type (WT) and *morA* mutant of *P. aeruginosa*. From the results of the expression studies, several genes that showed significant difference between WT and mutant were selected for further analysis. This led to the development of networks which linked genes based on similarities in their promoter or regulatory regions. Genes that were highly connected in the network were of great interest and further studies on these genes may help to shed light into the relationship between *P. aeruginosa* motility and its virulence

1. Introduction

The ability to move confers numerous advantages to a bacterium, including movement towards favourable conditions or avoidance of detrimental conditions, and successful competition with other microorganisms (Fenchel, 2002). In pathogenic bacteria, motility is usually considered a virulence factor essential for colonization of host organism or target organ (Ottemann *et al.*, 1997; Josenhans *et al.*, 2002).

We have recently described a novel membrane-bound regulator, MorA, that controls the timing of flagella development and its loss leads to changes in motility, chemotaxis

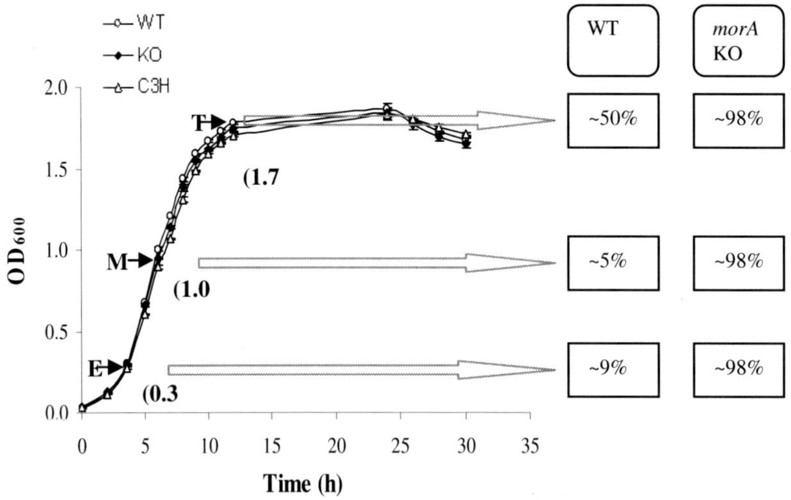

Figure 1. Timing of flagellar development in *P. putida* regulated by *morA*. Growth curves of *P. putida* WT, KO and C3H in LB medium showed no differences in the growth of the three strains. Cultures were sampled at early-log phase (0.3 OD_{600}, E), mid-log phase (1.0 OD_{600}, M) and log-to-stationary transition phase (1.7 OD_{600}, T) for TEM studies. Values are based on the average of two independent experiments with three replicates each. Standard errors are represented as vertical bars. Proportion of flagellated cells was expressed as a percentage of the total number of WT and KO cells counted using Transmission electron microscopy. Counts were based on an average of 400 cells. (Diagram and description extracted from Choy *et al.*, 2004).

and biofilm formation in *Pseudomonas* without affecting growth rate or cell size. MorA structure possesses a transmembrane domain, the redox-sensing PAS-PAC domain and the catalytic GGDEF and EAL domains, which are implicated in regulating the 2nd messenger cyclic-di-GMP (c-diGMP) levels. In the soil bacterium *Pseudomonas putida*, the loss of MorA derepresses flagellar development leading to flagella formation and increased motility in all the growth phases (Figure 1). The biofilm formation ability of *morA* knockout (KO) mutants is also reduced in diverse species such as *P. putida* and the human pathogen *P. aeruginosa*.

Since our first report (Choy *et al.*, 2004), we studied the effects of MorA on gene expression using Affymetrix full genome microarrays (*P. aeruginosa* Genome a GeneChip) in the planktonic phase of *P. aeruginosa* PAO1 and its *morA* mutant. Although there was no observable change in the motility of *P. aeruginosa* upon *morA* mutation, we were interested in identifying genes that were affected by the loss of *morA* during the planktonic phase, which is prior to the transition into biofilms.

2. Methods and Discussion

Total RNA was harvested from the early growth phase of *P. aeruginosa* PAO1 and its *morA* mutant strains. The total RNA was first treated with DNaseI and then purified with RNeasy Mini Kit (QIAGEN). cDNA was synthesized using the Superscript^{TM} First-Strand Synthesis System for RT-PCR (Invitrogen). The cDNA was fragmented using DNaseI and the fragmented product was labelled with biotin-ddUTP using BioArray^{TM} Terminal Labeling Kit (Enzo Life Sciences, Inc., Farmingdale, NY, USA) as per manufacturer's instructions. A gel-shift assay was then performed to ascertain whether the fragmented cDNA was properly labelled. The biotin-ddUTP labeled fragmented cDNA were then hybridized to the *P. aeruginosa* genome array GeneChip as per manufacturer's instructions (Affymetrix^{TM}, Inc., Santa Clara, CA, USA).

Three independent repeats of the experiment were done, each with WT and *morA* mutant comparisons. A large number of genes were found to be affected from the microarray and was selected for further analysis based on the following criteria: (i) they had more than 2-fold changes; (ii) they were assigned the "P" call value from analysis using the Affymetrix data mining software and (iii) they were statistically different using the Student's t-tests ($P \leq 0.05$).

Groups of genes that pass the 3 criteria stated above were split into up- or down-regulated and were hierarchically clustered using "TIGR Multiple Array Viewer". Different levels of the genes in the clusters were recorded and their gene numbers were extracted.

The extracted gene numbers were sent for promoter analysis where their promoters were extracted. The extracted promoters' motifs were analyzed by motif search algorithm which uses three heuristic algorithms: Tabu Search (TS), Simulated Annealing (SA) and a population-based Genetic Algorithm (GA) (Yang *et al.*, 2005). These were used for computational extraction of conserved motifs from a set of unaligned DNA sequences (promoters). The reason to use heuristics is to ensure that a reasonable, near-optimal solution to the optimization problem in question can be found in an acceptable time (Yang *et al.*, 2005). This tool can be assessed freely at http://sdmc.i2r.a-star.edu.sg/DRAGON/Motif_Search/ for academic and non-profit users.

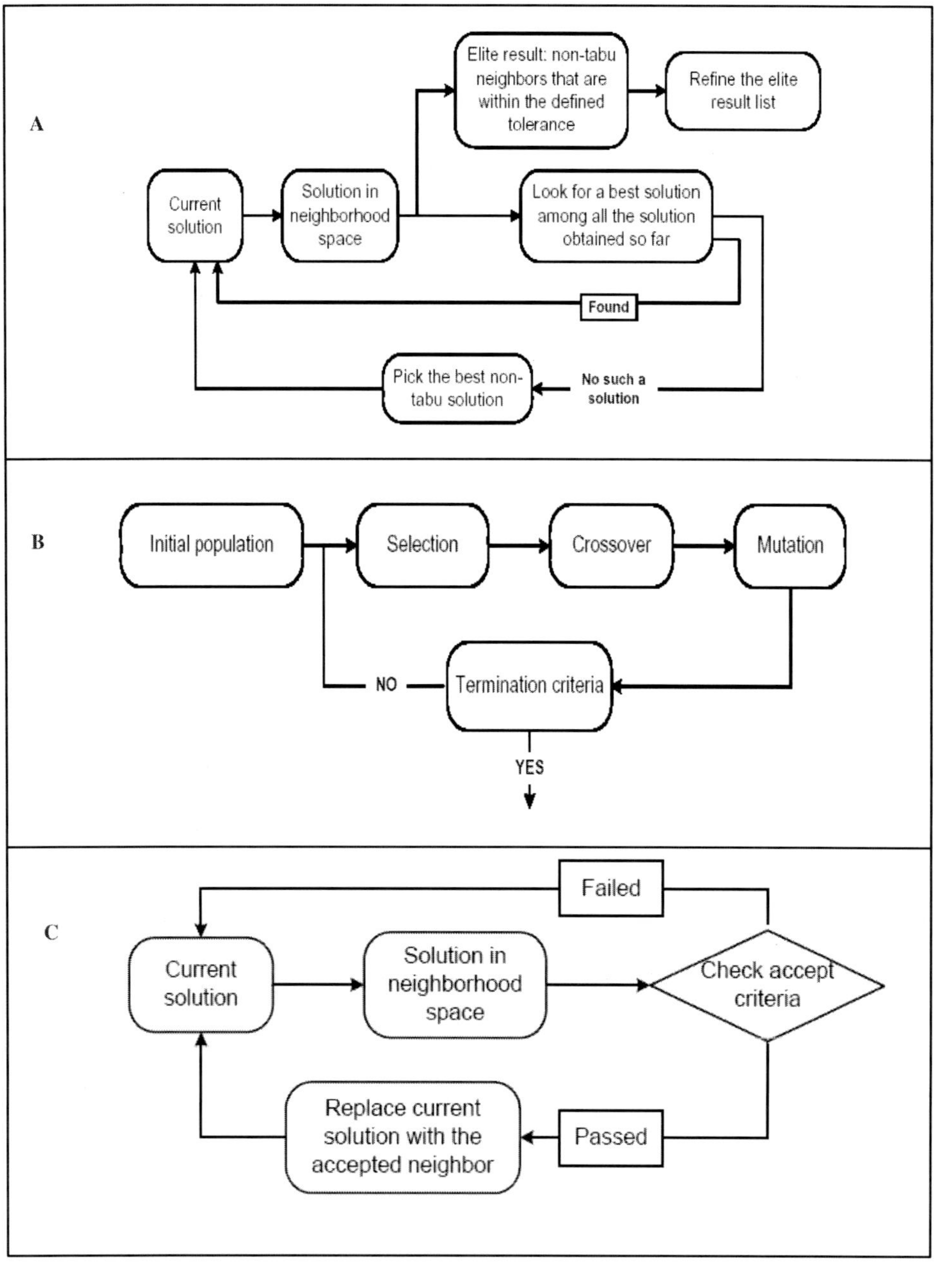

Figure 2. Flowchart of data processing within TS (A), GA (B) and SA (C).

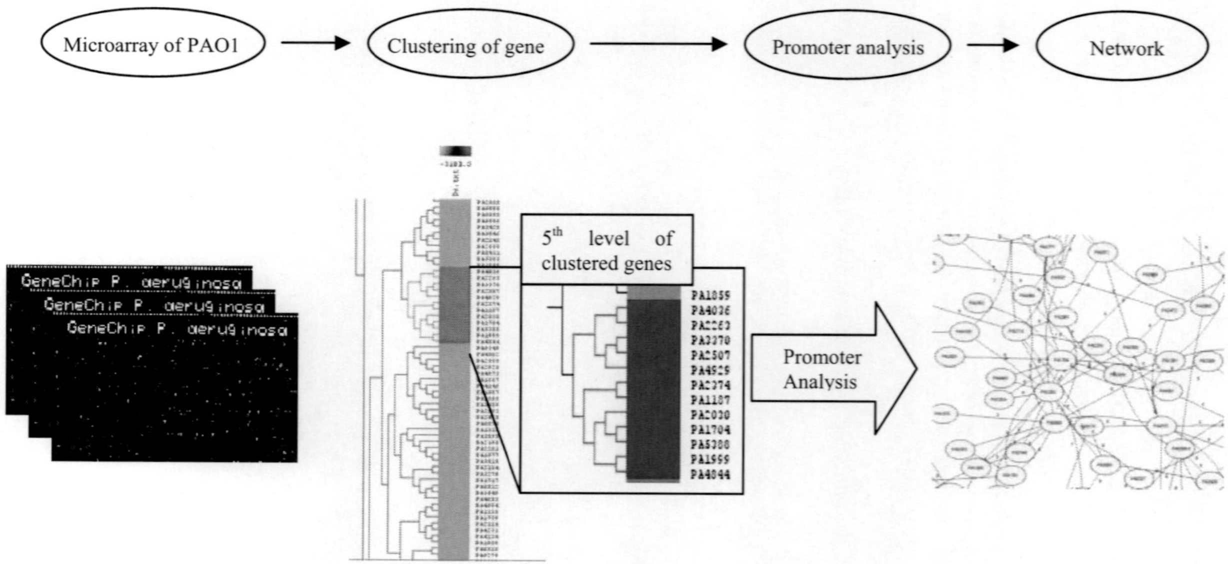

Figure 3. Flowchart showing the process from raw data to networks. Initial work started by the studying of gene expression using total RNA samples from *P. aeruginosa* to perform microarray (Affymetrix). The microarray results were statistically analyzed and genes that passed the t-tests were hierarchically clustered. The Gene IDs from different levels of the clustered genes were extracted and sent for promoter analysis. Lastly, a network is obtained showing the relationship between different genes promoters' motifs.

To prevent or penalize the selection of solutions from the points in the solution space, TS was used to avoid remaining in iteration cycles by using memory techniques while SA is suitable for discrete combinatorial optimization problems by using the principle of thermodynamics. By using selection, crossover and mutation as its three fundamental principles, GA is able to create new population. It uses a number of current solutions and combines them together to generate new solutions by imitating the genetic process of reproduction (Yang *et al.*, 2005). The drawback of these algorithms is lower speed, but consistency of the extracted pattern groups is usually considerably higher than what is obtained with expectation maximization or with Gibbs sampling (Yang *et al.*, 2005). Figure 2 show the flowchart of data processing within TS, GA and SA.

The relationships between the promoters were studied and regulatory networks were generated using an in-house networking program. From the networks, we were able to distinguish which genes may have shared motifs in their promoter. Genes that were well connected to other genes may be an indication that its up- or down-regulation will affect the other genes that were connected to it (a summary of the flow of work is shown in Figure 3.

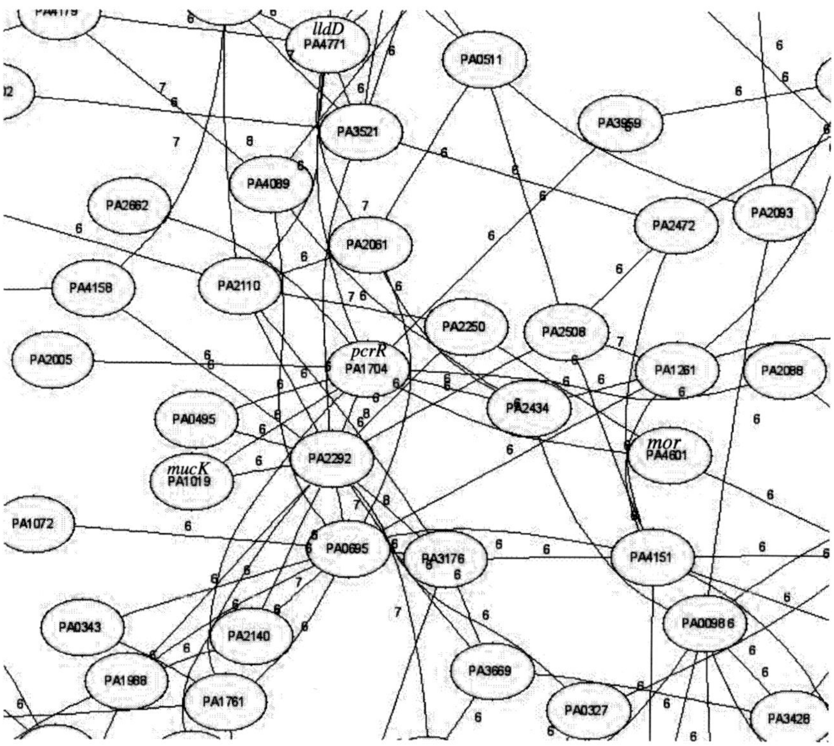

Figure 4. Network of PA1704 (*pcrR*). The figure was obtained from a cluster of genes showing down-regulation in *P. aeruginosa*. The gene of interest here is *pcrR* as it is highly connected to other genes especially morA, which is our gene of study.

3. Conclusion

From these analyses, we identified several genes with shared promoter motifs (Figure 4). Among the nodes of high density of connections, one prominent gene was *pcrR*. It is highly connected to 11 other genes together with *morA*. *pcrR* codes for a transcriptional regulator and controls bacterial virulence via Type III Secretion. Thus, we hypothesize that *morA* may also regulate virulence genes via Type III Secretion. Further studies are being conducted to validate the relationship between MorA and PcrR.

Acknowledgments

This work was supported by grants R-154-000-145-112, R-154-000-114-112 and R154-000-253-112 from the Academic Research Fund, National University of Singapore. Weng-Keong Choy was funded by a research scholarship awarded by the National University of Singapore.

References

1. Bonfiglio G., Carciottoa V., Russoa G., Stefania S., Debbiab G.C.S.E. and Nicolettia G. Antibiotic resistance in *Pseudomonas aeruginosa*: an Italian survey. *Journal of Antimicrobial Chemotherapy* 41:307–310, 1998.
2. Choy W.K., Zhou L., Chris Syn K. C., Zhang L.H., and Swarup S. MorA Defines a New Class of Regulators Affecting Flagellar Development and Biofilm Formation in Diverse *Pseudomonas* Species. *Journal of Bacteriology* 186:7221–7228, 2004.
3. Fenchel T. Microbial behavior in a heterogeneous world. *Science* 296:1068-1071, 2002.
4. Josenhans, C. and Suerbaum, S. The role of motility as a virulence factor in bacteria. *Int. J. Med. Microbiol.* 291:605-614, 2002.
5. Ottemann, K.M. and Miller, J.F. Roles for motility in bacterial-host interactions. *Mol. Microbiol.* 24:1109-1117, 1997.
6. Pitt T.L., Sparrow M., Warner M. and Stefanidou M. Survey of resistance of *Pseudomonas aeruginosa* from UK patients with cystic fibrosis to six commonly prescribed antimicrobial agents. *Thorax* 58:794-796, 2003.
7. Quinn, J.P. Clinical problems posed by multiresistant nonfermenting gram-negative pathogens. *Clinical Infectious Diseases* 27:S117-S124, 1998.
8. Stover C.K., Pham X.Q., Erwin A.L., Mizoguchi S.D., Warrener P., Hickey M.J., Brinkman F.S., Hufnagle W.O., Kowalik D.J., Lagrou M., Garber R.L., Goltry L., Tolentino E., Westbrock-Wadman S., Yuan Y., Brody L.L., Coulter S.N., Folger K.R., Kas A., Larbig K., Lim R., Smith K., Spencer D., Wong G.K., Wu Z, Paulsen I.T., Reizer J., Saier M.H., Hancock R.E., Lory, S. and Olson M.V. Complete genome sequence of *Pseudomonas aeruginosa* PA01, an opportunistic pathogen. *Nature* 406:959-964, 2000.
9. Yang L., Huang E. and Bajic V.B. Some Implementation Issues of Heuristic Methods for Motif Extraction From DNA Sequences. *International Journal of Computers, Systems and Signals* 6:3-12, 2005.

AUTHOR INDEX